Laboratory Anatomy of the
Perch

Fourth Edition
Laboratory Anatomy of the

Perch

Robert B. Chiasson
University of Arizona

William J. Radke
Central State University

 Wm. C. Brown Publishers

Laboratory Anatomy Series

Consulting Editor
Robert B. Chiasson

Laboratory Anatomy of the Cat
Robert B. Chiasson
(Ernest S. Booth)

Laboratory Anatomy of the Fetal Pig
Theron O. Odlaug

Laboratory Anatomy of the Frog
Raymond A. Underhill

Laboratory Anatomy of the Human Body
Bernard B. Butterworth

Laboratory Anatomy of the Mink
David Klingener

Laboratory Anatomy of the Perch
Robert B. Chiasson
William J. Radke

Laboratory Anatomy of the Pigeon
Robert B. Chiasson

Laboratory Anatomy of the Rabbit
Charles A. McLaughlin
Robert B. Chiasson

Laboratory Anatomy of the Shark
Laurence M. Ashley
Robert B. Chiasson

Laboratory Anatomy of the Turtle
Laurence M. Ashley

Laboratory Anatomy of the White Rat
Robert B. Chiasson

Illustrated by the Authors

Book Team

Editor *Kevin Kane*
Developmental Editor *Margaret J. Manders*
Production Coordinator *Carla D. Arnold*

Wm. C. Brown Publishers

President *G. Franklin Lewis*
Vice President, Publisher *George Wm. Bergquist*
Vice President, Publisher *Thomas E. Doran*
Vice President, Operations and Production *Beverly Kolz*
National Sales Manager *Virginia S. Moffat*
Advertising Manager *Ann M. Knepper*
Marketing Manager *Craig S. Marty*
Editor in Chief *Edward G. Jaffe*
Managing Editor, Production *Colleen A. Yonda*
Production Editorial Manager *Julie A. Kennedy*
Production Editorial Manager *Ann Fuerste*
Publishing Services Manager *Karen J. Slaght*
Manager of Visuals and Design *Faye M. Schilling*

Cover design by Sailer & Cook Creative Services

Some of the laboratory experiments included in this text may be hazardous if materials are handled improperly or if procedures are conducted incorrectly. Safety precautions are necessary when you are working with chemicals, glass test tubes, hot water baths, sharp instruments, and the like, or for any procedures that generally require caution. Your school may have set regulations regarding safety procedures that your instructor will explain to you. Should you have any problems with materials or procedures, please ask your instructor for help.

Copyright © 1966, 1974, 1980, 1991 by Wm. C. Brown Publishers. All rights reserved

Library of Congress Catalog Card Number: 90-81805

ISBN 0-697-04939-6

No part of this publication may be reproduced, stored in a retrieval system, or transmitted, in any form or by any means, electronic, mechanical, photocopying, recording, or otherwise, without the prior written permission of the publisher.

Printed in the United States of America by Wm. C. Brown Publishers, 2460 Kerper Boulevard, Dubuque, IA 52001

10 9 8 7 6 5 4 3 2 1

Contents

List of Figures **vii**

Preface **ix**

1 Introduction **1**

2 External Anatomy and Skin **7**

3 The Skeletal System **14**

4 The Muscular System **27**

5 Mouth, Pharynx, and Respiratory System **40**

6 Body Cavities and Viscera **43**

7 The Digestive System **46**

8 The Circulatory System **52**

9 Excretory and Reproductive Systems **62**

10 The Nervous System **68**

11 The Endocrine System **80**

Index **85**

List of Figures

1.1	Phylogenetic chart of the vertebrates **2**	4.5	Opercular musculature **33**
1.2	Distribution of *Perca flavescens* **4**	4.6	Jaw musculature **34**
1.3	Planes and directions of the perch body **5**	4.7	Lateral view of appendicular musculature **36**
2.1	External anatomy of the perch **8**	4.8	Ventral view of jaw musculature **37**
2.2	Body form and locomotion **9**	4.9	Musculature of the median fin **37**
2.3	Types of tails found in fishes **10**	4.10	Musculature of the tail **38**
2.4	Semidiagrammatic cross section of the skin **12**	4.11	Caudal view of the left eyeball **38**
2.5	Ctenoid scale of the perch **12**	5.1	Oral cavity and pharynx **40**
2.6	A comparison of the ctenoid, cycloid, and placoid scales **13**	5.2	Diagram of the gills and gill circulation **41**
3.1	Lateral view of the skull **15**	6.1	Right lateral view of the visceral organs **44**
3.2	Dorsal view of the skull **16**	7.1	Diagrammatic right lateral view of the intestinal tract **46**
3.3	Medial view of the right jaws **17**	7.2	Diagram of a hypothetical cross section of the fish intestinal tract illustrating features of several regions of the tract **47**
3.4	Ventral view of the skull with lower jaws, hyoid, and branchial arches removed **18**	7.3	Photomicrographs of cross sections of portions of the corpus and pyloric regions of the stomach and of the small intestine $\times 100$ **48**
3.5	Caudal view of the skull and pectoral girdle **18**	7.4	Photomicrograph of a portion of a cross section of the fish intestine $\times 300$ **49**
3.6	Lateral view of the hyoid apparatus **20**	7.5	Photomicrograph of a portion of the fish liver and pancreas $\times 500$ **50**
3.7	Dorsal view of the branchial apparatus **20**	7.6	Drawing of an islet of Langerhans from a fish pancreas $\times 1000$ **50**
3.8	Lateral view of the entire skeleton **21**	8.1	The heart and associated vessels **53**
3.9	Representative vertebrae **22**	8.2	Branchial circulation **54**
3.10	Caudal vertebrae and tail, lateral view **24**	8.3	Visceral organs and blood vessels on the right side **56**
3.11	Lateral view of the left pectoral girdle and limb **25**	8.4	Visceral organs and blood vessels on the left side **57**
3.12	Pelvic girdle and fins **25**	8.5	Segmental branches of the dorsal aorta **58**
4.1	Lateral view of the musculature **28**		
4.2	Cross section of the trunk and tail regions **29**		
4.3	Diagram of myomere movement **29**		
4.4	Diagram illustrating the gradation of body constriction when slanted myotomes are innervated **30**		

8.6	Caudal cardinal veins and kidneys **59**	10.1	Lateral view of the brain **69**
8.7	Diagram of the main cardinal drainage system **60**	10.2	Dorsal and ventral views of the brain **69**
8.8	Section of the fish spleen ×100 **61**	10.3	Midsagittal view of the brain **70**
9.1	Drawing of (A) fish kidney cross sectioned approximately half way between anterior and posterior extremes and (B) a diagram of a single fish kidney tubule **62**	10.4	Cross section of the perch midbrain (mesencephalon) **72**
		10.5	Spinal cord and the arrangement of spinal nerves **74**
9.2	Details of a glomerulus (A) and kidney tubules (B, C, D, and E) **63**	10.6	The central nervous system and appendicular plexus **75**
9.3	Lateral view of the male urogenital organs **65**	10.7	Diagram of the lateral line organ **76**
		10.8	Relationship of the eyes and ears to the brain **77**
9.4	Dorsal views of the male reproductive system **66**	10.9	Lateral view of the inner ear **78**
9.5	Semidiagrammatic cross section of the ovary and associated membranes **67**	10.10	Sagittal view of the eye **78**
		11.1	Endocrine glands **81**

Preface

Although the Yellow Perch (*Perca flavescens*) has long been used as a dissection specimen for laboratory courses, there is no complete description of the anatomy of the perch in the English language. Many elementary zoology and comparative anatomy laboratory manuals treat the perch superficially, but the dogfish shark is usually considered as the "fish" to be studied in detail. While it is true that the dogfish is easily dissected, it is not representative of the majority of fishes nor of a "primitive" fish. Indeed, the dogfish is a highly specialized form quite distinct from all bony fishes. The perch is representative of the largest vertebrate class, Osteichthyes.

Thus, in spite of the fact that bony fishes make up the largest living class of vertebrates, their anatomy is virtually ignored or, at best, glossed over by the authors of textbooks and laboratory manuals. Hopefully, this manual will help to fill some of the void in this area.

In addition to a systematic description of the anatomy of the perch, certain structures of other fishes are included in this manual for comparison. The other fishes that are included are available from biological supply houses and help to show the diversity of form in the vast group of vertebrates.

As in the other manuals of this series, more descriptive material is included than is usually covered in an undergraduate course. This provides some latitude of choice on the part of the laboratory instructor.

DISSECTION TECHNIQUES

Some special materials will be necessary if all of the dissections and investigations described in this manual are to be covered.

Microscopic slides of the skin and of entire scales are available from biological supply firms. The study of these slides will require a compound microscope with magnification to 400×. Investigation of some of the dissections (brain, for example) may require a dissecting microscope.

Prepared skeletal material is available from most biological supply firms. Generally this material is not thoroughly cleaned and unless the instructor carefully cleans the skeletons the student will not be able to see many of the skull bones. If a thorough dissection is desired, it might be best to have an early cursory examination of the prepared material and a later, complete study of the skull of the student's dissection specimen. In such case, the second study of the skull should precede the study of the nervous system but should follow the study of the muscular, respiratory, and circulatory systems.

It is assumed that injected specimens will be available for study of the circulatory system. Specimens with four separate injections, (1) prebranchial arteries; (2) postbranchial arteries; (3) caudal vena cava; and (4) hepatic portal venous, will be helpful for demonstration, if they can be obtained.

Skinning is usually a simple procedure. However, connective tissue of the skin may adhere to the underlying muscles and may tear the muscles as you pull the skin away. Extreme care should be taken to avoid destruction of the small fin muscles during skinning. A study of the head muscles may require removal of the opercular shield and a portion of the lower jaw. These bones should be removed from one side only. The latter restriction will preserve the opposite side for later study.

Additional specific directions for dissections will be found at appropriate places in the text.

PRONUNCIATION KEY

A pronunciation guide is provided for the terminology used in this manual. Each term is spelled phonetically immediately after it first appears in the text. The following general rules were used in creating the guide.

1. The syllable with the strongest accent appears in capital letters, for example: a-NAT-*o*-me.
2. Vowels to be pronounced in the long form are in italics, as in: bl*a*de and b*i*te.
3. Unmarked vowels are pronounced in the short form, as in: mitt and drum.
4. Other indicators of sounds are: oo as in blue, yoo as in cute, oy as in foil.

The student will find learning the correct spelling of a term easier if just a few moments are spent studying the proper pronunciation.

ACKNOWLEDGMENTS

We wish to thank the reviewers and users of this manual. Their comments concerning the retention, deletion, or alteration of material as well as the correction of errors has been most valuable. We wish to thank Dr. Lawrence M. Page, Illinois State Natural History Survey Division for permission to use the reprinted painting of the perch from "The Fishes of Illinois" (1908) by S. A. Forbes and R. E. Richardson. We also thank John Wiley & Sons, publishers of *Fishes of the World* (2nd ed., 1984) by J. S. Nelson for permission to reprint the classification of fishes from that work. We are grateful to David L. Bentley for help in the preparation of the index.

Chapter 1
Introduction

SYSTEMATIC POSITION

Figure 1.1 illustrates the phylogenetic relationships, relative abundance (in terms of number of species), and historical occurrence of vertebrates. Bony fishes are the largest group of vertebrate animals living today, both in number of individuals and in number of species (about 30,000). They occur in fresh and sea water, from the North Pole to the Antarctic, and completely around the world. They live at depths as great as seven miles in the oceans, as high as three miles above sea level in mountain streams or lakes.

Such great diversity has produced many problems in the classification of fishes and even more problems to those workers attempting to discover the evolutionary interrelationships of fishes. These problems have led to considerable argument among ichthyologists as to the classification and evolution of fishes.

The classification of fishes has been reconsidered by Nelson, *Fishes of the World* (2nd ed. 1984) and that classification is followed in this revision.

The order Perciformes (PERK-i-for-meez) is the largest order of fishes in terms of the number of families (150), genera (1367) and species (7791). Perciformes are the most diverse of all fish orders and are dominate forms in both marine (75% of the species) and freshwater (25% of the species) habitats. Two suborders (Percoidei and Gobioidei) make up 66 percent of all Perciform species. Percoidei (perk-OY-de-i) is the major suborder with about 3524 species, many of which are desirable as human food including striped bass, bluefish, snappers, barracudas, sunfishes, and perches. The family Percidae (PERK-i-da) contains all of the freshwater perches and all are found in the northern hemisphere as nine genera with 146 species. Ninety percent occur in North America east of the Rocky Mountains and are mostly darters. The Percidae have two dorsal fins, one or two anal spines, and pelvic fins that are located in the thoracic region. The vertebrae number 32 to 50 with a maximum body size of 90 cm attained by the walleye *Stizostedion vitreum*. For more information regarding the Percids see vol. 34(10) of the *Journal of the Fisheries Research Board of Canada* (1977). The genus *Perca* (PERK-a) has three species: *P. fluviatilis,* a Eurasian species, *P. flavescens* (fla-VES-ens) of North America, and *P. shrenki* in Asia. All three are generalized forms that probably represent the ancestral type from which the other species were derived. This is speculation of course but most aspects of the anatomy of the perch are widely accepted as typical of the Teleostei (tel-e-OS-te-i).

The following classification follows J. S. Nelson (1984) *Fishes of the World*. It lists all of the orders of living teleost fishes. The extinct orders are omitted.

Phylum Chordata

    Superclass Agnatha (jawless vertebrates)
        Class Myxini (hagfishes)
        Class Cephalaspidomorphi (lampreys)
    Superclass Gnathostomata (jawed vertebrates)
        Class Chondrichthyes (cartilaginous fishes)
            Subclass Elasmobranchii (sharks, rays, skates)
            Subclass Holocephali (chimaeras)
        Class Osteichthyes (bony fishes)
            Subclass Dipneusti (lungfishes)
            Subclass Crossopterygii (fringe-finned fishes)
            Subclass Brachiopterygii (bichirs)
            Subclass Actinopterygii (ray-finned fishes)
                Infraclass Chondrostei (sturgeons)
                Infraclass Neopterygii

J. S. Nelson, *Fishes of the World*. Copyright © John Wiley and Sons, Inc., 1984. Reprinted by permission of John Wiley and Sons, Inc.

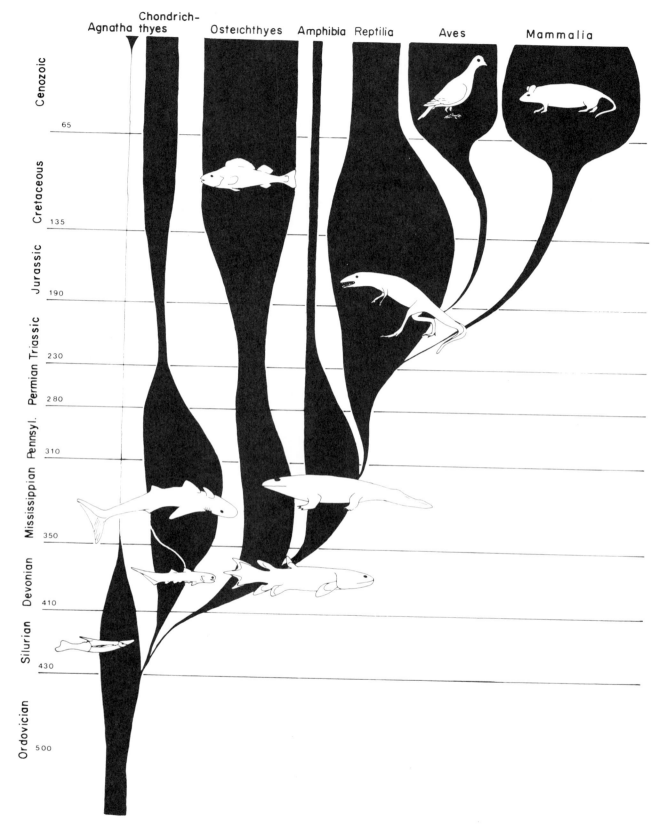

Figure 1.1 Phylogenetic chart of the vertebrates. The name of the vertebrate class is at the top of the chart. The name of the geological period and the approximate time before the present of the geological period is presented on the left of the chart. Time is presented in millions of years. The class Placoderma is unlabeled but is represented as a segment of the class Chondrichthyes. Ostracoderms are included with the Agnatha and the Acanthodians are included with the Osteichthyes.

Division Ginglymodi (gars)
Division Halecostomi
  Subdivision Halecomorphi (bowfin)
  Subdivision Teleostei (teleosts)
    Infradivision Osteoglossomorpha
      Order Osteoglossiformes (bonytongues, butterflyfish, mooneyes)
    Infradivision Elopomorpha
      Order Elopiformes (tarpons, bonefishes)
      Order Anguilliformes (typical eels)
      Order Notacanthiformes (deep-sea eels)
    Infradivision Clupeomorpha
      Order Clupeiformes (herrings)
    Infradivision Euteleostei
     Superorder Ostariophysi
      Order Gonorynchiformes (milkfish)
      Order Cypriniformes (carps, suckers)
      Order Characiformes (hatchetfishes)
      Order Siluriformes (catfishes)
      Order Gymnotiformes (knifefishes)
     Superorder Protacanthopterygii
      Order Salmoniformes (pikes, smelts, salmon)
     Superorder Stenopterygii
      Order Stomiiformes (lightfishes, dragonfishes)
     Superorder Scopelomorpha
      Order Aulopiformes (greeneyes, lizardfishes)
      Order Myctophiformes (lanternfishes)
     Superorder Paracanthopterygii
      Order Percopsiformes (trout-perches, cavefish)
      Order Gadiformes (cods)
      Order Ophidiiformes (cusk-eels)
      Order Batracoidiformes (toadfishes)
      Order Lophiiformes (anglerfishes)
      Order Gobiesociformes (clingfishes)
     Superorder Acanthopterygii
      Order Cyprinodontiformes (flyingfishes, livebearers, guppy)
      Order Atheriniformes (grunion)
      Order Lampriformes (ribbonfishes)
      Order Beryciformes (squirrelfishes)
      Order Zeiformes (dories)
      Order Gasterostiformes (sticklebacks)
      Order Indostomiformes (*I. paradoxus*)
      Order Pegasiformes (seamoths)
      Order Syngnathiformes (pipefishes, seahorses)
      Order Dactylopteriformes (flying gurnards)
      Order Synbranchiformes (swamp-eels)
      Order Scorpaeniformes (sculpins)
      Order Perciformes (basses, sunfishes, perches (includes *Perca flavescens*), bluefishes, remoras, dolphins, snappers, drums, ciclids, mullets, eelpouts, blennies, gobies, mackerels, tunas, swordfish)
      Order Pleuronectiformes (flounders, soles)
      Order Tetraodontiformes (puffers, molas)

The classification of the Yellow Perch according to Nelson (1984) is as follows:

Phylum: Chordata (k*or*-DA-ta)
  Subphylum: Vertebrata (VER-te-bra-ta)
    Superclass: Gnathostomata (na-TH*OS*-t*o*-ma-ta)
      Class: Osteichthyes (os-t*e*-ICK-th*ee-ee*z)
        Subclass: Actinopterygii (ak-ti-nop-te-RIJ-*i*)
          Infraclass: Neopterygii (n*e*-op-te-RIJ-*i*)
            Division: Halecostomi
              Subdivision: Teleostei (t*e*l-*e*-OS-t*e*-*i*)
                Infradivision: Euteleostei (*u*-tel-*e*-OS-t*e*-*i*)
                  Superorder: Acanthopterygii (ak-an-thop-ter-IJ-*e*-*i*)
                    Order: Perciformes (PERK-i-f*or*-m*ee*z) (pelvic fin originates anterior to pectoral)
                      Suborder: Percoidei (perk-OY-d*e*-*i*) (73 families)
                        Superfamily: Percoidea (perk-OY-d*e*-a)
                          Family: Percidae (PERK-i-d*a*) (nine genera)
                            Subfamily: Percinae (PERK-i-n*e*) (freshwater, Northern Hemisphere)
                              Tribe: Percini (PERK- i-n*i*) (3 genera)
                                Genus: *Perca* (PERK-a)
Scientific name: *Perca flavescens* (fla-VES-ens)
Common Name: Yellow Perch

The European counterpart of the Yellow Perch is the European Perch, *Perca fluviatilis*. The two species are extremely similar and are separable only on minor differences. Comments in this manual may be applied to either species.

Figure 1.2 Distribution of *Perca flavescens*.

PLANES AND DIRECTIONS

Prior to the publication of *Nomina Anatomica Veterinaria* (1968) anatomists used *anterior* (an-TEER-e-or) to mean toward the head and *posterior* (pos-TEER-e-or) to indicate toward the buttock (or tail). This has been revised so that *cranial* (KRA-ne-al) is used to mean toward the head and *caudal* (KAW-dal) means toward the tail. Although there is no such guide for fish it seems reasonable to follow these same practices.

Within the head, the term *rostral* (ROS-tral) is used to indicate a forward direction. Rostral is really an imaginary point in front of the head and as used here means toward that point.

The more commonly used directions and planes are illustrated in figure 1.3. In addition to those mentioned above there are three sets of opposing terms that are illustrated and frequently used. *Medial* (Me-de-al) means toward the *sagittal* (SAJ-i-tal) plane in the center of the body and *lateral* (LAT-er-al) is away from the sagittal plane. *Dorsal* (DOR-sal) is above the *frontal* (FRUN-tal) plane and *ventral* (VEN-tral) is below the frontal plane. The terms *proximal* (PROK-si-mal) and *distal* (DIS-tal) are used in connection with appendages (fins)—toward the free end of the appendage is distal and toward the body is proximal.

In addition to the frontal and sagittal planes just mentioned, a third plane is the *transverse* (trans-VERS), which is also referred to as a *cross section*. This may section the animal at any level from rostral to caudal at right angles to the long axis.

Some terms may be formed by adding a prefix to the name of an anatomical feature. Some common prefixes are: *Pre-*, before or in front (this prefix usually refers to a position); *Post-*, behind or after in position; *Para-*, near or alongside; *Supra-*, above; and *infra-*, below a given structure.

SUGGESTED READINGS

Berg, L. S. 1940. Classification of fishes both recent and fossil. *Trav. Inst. Zool. Acad. Sci. URSS* 5:87–517. Reprinted 1947 by Edwards Brothers, Ann Arbor, Mich.

Grasse, P. 1958. Agnathes et poissons anatomie, ethologie, systematique. *In Traite de Zoologie*. Tome XIII, Mason et Cie Paris.

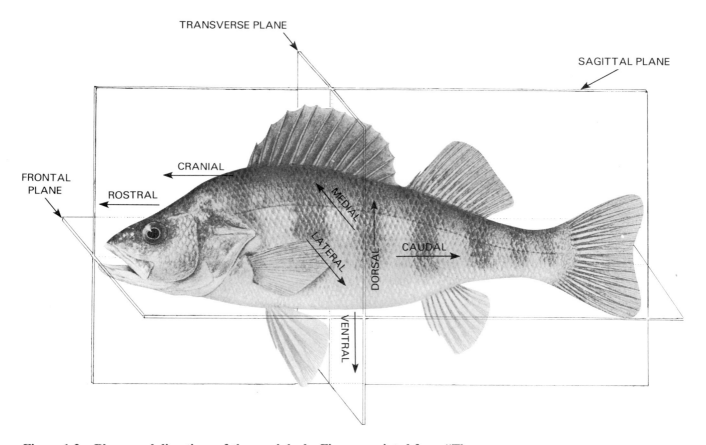

Figure 1.3 Planes and directions of the perch body. Figure reprinted from "The Fishes of Illinois." [1908. S. A. Forbes and R. E. Richardson. Illinois Natural History Survey.]

Gregory, W. K. 1933. Fish skulls: A study of the evolution of natural mechanisms. *Trans. Amer. Philos. Soc.* 23:75–481.

Jordan, D. S. 1923. A classification of fishes including families and genera as far as known. *Stanford Univ. Publ. Biol. Sci.* 3(2):79–243.

Lauder, G. V. and K. F. Liem. 1983. The evolution and interrelationships of the Actinopterygian fishes. *Bull. Mus. Comp. Zool.* 150(3):96–197.

Nelson, J. S. 1984. *Fishes of the world* (2nd ed.). New York and London: John Wiley & Sons.

Regan, C. T. 1929. Fishes. In *Encyclopedia Britannica* (14th ed.), 9:305–328.

Romer, A. S. 1966. *Vertebrate paleontology.* Chicago and London: University of Chicago Press.

Wiley, E. O. 1979. Ventral gill arch muscles and the interrelationships of gnathostomes with a new classification of the vertebrata. *Zool. Jour. Linn. Soc.* 67(2):149–180.

Chapter 2
External Anatomy and Skin

The body of a fish is divided into three parts—*head, trunk,* and *tail* (fig. 2.1). The head and trunk are separated from each other by the gill openings which are covered laterally by a bony *operculum* (*o*-PER-k*u*-lem). The caudal edge of the operculum provides a good, practical, caudal border for the head. The caudal limit of the trunk in the perch is marked by the ventral, *anal* (*A*-nal), and *urogenital* (yoo-ro-JEN-i-tal) openings.

The most obvious features of the fish's appearance are (1) the streamlined shape of the body, and (2) the fins (paired and unpaired) that break the otherwise smooth contours of the fish's body. These features are adaptations for swimming and provide the animal with advantages over its prey.

BODY SHAPE

Figure 2.2 is a dorsal view of a swimming perch and representative cross sections (A through C) illustrating the cranial-caudal changes in body shape. The snout and head are a square outline (A) but this changes in the area of the trunk to a roughly "teardrop" shape (B) and is a laterally compressed to an oval shape through the caudal peduncle (p*e*-DUNG-kl) (C).

There are several different categories for swimming based on body motions (or lack of) and the use (or non-use) of the caudal fin. Some fish use the paired fins in "rowing" movements and still others may use the dorsal (the "Bowfin," *Amia*) or ventral fin (*Gymnotus,* an electric fish without a caudal fin) for propulsion. Most fish use the body and caudal fin with sinuous, eel-like body motions (*Anguilliform*) or a more limited body "wave" movement (*Carangiform* or *Subcarangiform*). The swimming form of small sharks is anguilliform (named for the eel, *Anguilla*) which may be described as sinuous contortions of the body. The perch produces its major propulsive force with its tail. As the tail bends laterally a thrust is produced against the water and it is this thrust that propels the animal's body forward, assisted by movements of the heterocercal tail (see figs. 2.2 and 2.3 and discussion of caudal fin that follows). The total force (TF) may be divided into two components, a lateral force (Fl) and a forward component (Ff). The lateral components of the several waves present on the body at any one time tend to cancel one another. The forward components are additive and propel the fish forward in the water.

THE HEAD

The head of the perch has three notable structures: a *mouth,* four *nasal apertures* (N*A*-zal AP-er-churz), and two *eyes.* The overall shape of the perch is *fusiform* (FU-zi-form) or torpedo-shaped. Consequently, the head is smaller than the trunk. There is a taper (in lateral view) from the mouth to the deepest part of the trunk, just behind the appendages. The slope is more abrupt on the dorsal side than it is on the ventral side.

The mouth of the perch is terminal rostrally but in some other fishes it may be dorsal or ventral. The position of the mouth on the head reflects the manner in which the fish feeds. If the mouth is dorsal, the fish is usually a surface feeder. If the mouth is ventral, the fish is usually a bottom feeder. If the mouth is terminal as in the perch, the fish usually feeds by overtaking prey while swimming.

The *lips* of the perch are membranous. Teeth are carried on the *dentary* (DEN-tar-*e*) of the *mandible* (MAN-di-bl) (lower jaw) and on the *premaxilla* (pre-MAK-sil-a) of the upper jaw. The *maxillae,* which are excluded from the gape in the perch, are toothless (see chapter 3).

There are four nasal apertures in the perch—two on each side of the midline just above the mouth. Water enters each *nasal sac* through the rostral nasal aperture, passes over the *olfactory epithelium* (*o*l-FAK-to-re ep-i-THE-le-um) (see page 79) and leaves by the caudal aperture. Lidless eyes are situated on each side of the perch head so that binocular vision is impossible.

THE TRUNK

The trunk increases in size caudally from the head to the middle of the *first dorsal fin.* Caudally, the tail tapers down from the trunk to the *caudal fin.* In cross section the trunk is "teardrop" shaped. This form greatly reduces drag against the water and conserves energy during locomotion (see fig. 2.2).

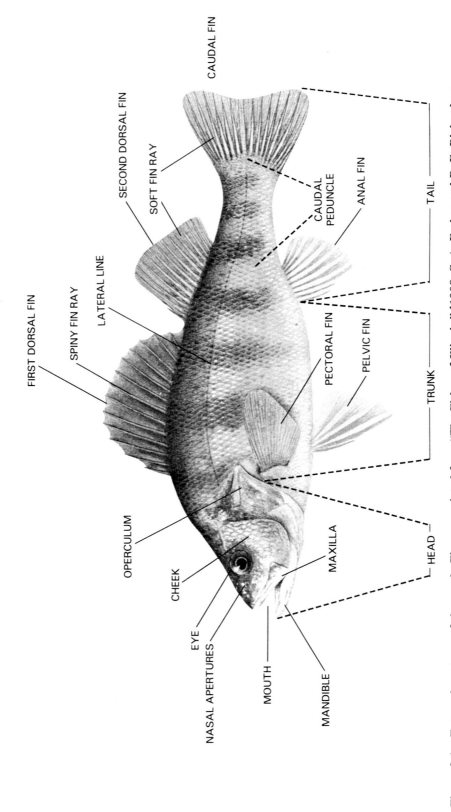

Figure 2.1 External anatomy of the perch. Figure reprinted from "The Fishes of Illinois." [1908. S. A. Forbes and R. E. Richardson. Illinois Natural History Survey.]

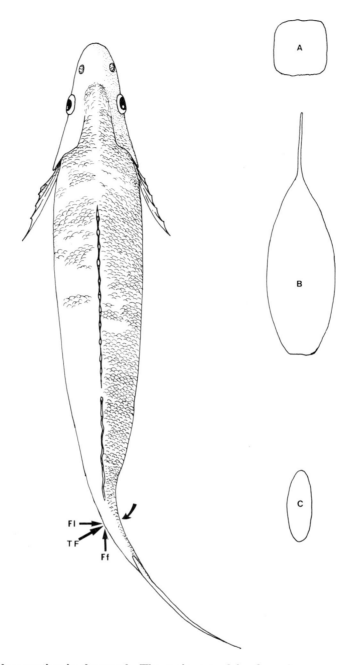

Figure 2.2 Body form and locomotion in the perch. The main propulsive force is generated by lateral movements of the tail, which is opposed by the resistance of the water producing a force in opposition to the body movement (curved arrow). The total force (TF) may be divided into two vectors representing a lateral component (Fl) and a longitudinal component (Ff). The propulsive longitudinal component is opposed by frictional drag that is greatly reduced by the *streamlining* contour of the body.

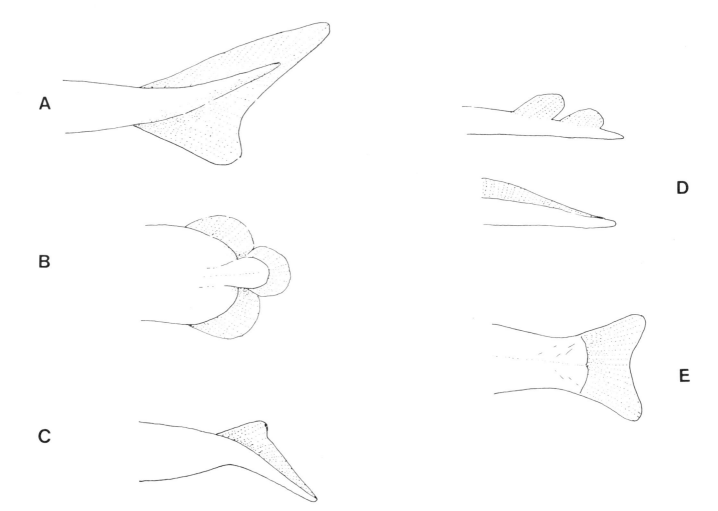

Figure 2.3 Types of tails found in fishes. The heterocercal tail (A) is typical of sharks and is correlated with a horizontal arrangement of the pectoral fins (see text, p. 11). Crossopterygian fishes (including the ancestor of tetrapods) have a diphycercal tail (B). The extinct anaspids (Ostracoderms; see fig. 1.1) had a hypocercal tail (C) and the perch have a homocercal tail (E). Electric fishes such as the elasmobranch, *Raja* (Rayfish) or the teleost, *Gymnarchus* (electric eel) have tail musculature modified for producing an electrical charge and these muscles are not functional in locomotion. Consequently the tails have no caudal fins (D).

Another term for describing the perch body (especially in cross section) is *compressed*. There are four categories of fish body shapes—*compressed* (flattened laterally), *depressed* (flattened dorso-ventrally), *truncated* (shortened cranio-caudally), and *attenuated* (thin or eel-like).

THE PAIRED APPENDAGES

The perch has two sets of paired appendages, the *pectoral* (PEK-to-ral) and *pelvic* (PEL-vik) fins. In spite of the fact that the two sets of paired fins are at the cranial end of the perch trunk, they are homologous to the paired appendages of tetrapods.

The paired pectoral fins are located high on the sides of the trunk and just caudal to the gill openings. In fact, the bones on which the pectoral fins are based are attached directly to the skull (see chap. 3).

The pelvic fins are set far cranial on the trunk, just caudal to the pectoral fins. The two pelvic fins are close together on the cranial ventral surface of the trunk in the thoracic position, as compared to the caudal abdominal pelvics of more primitive bony fishes.

THE UNPAIRED FINS

In addition to paired fins the perch has four median unpaired fins. The names of these fins indicate their position on the body. The *first dorsal fin* begins at the cranial-most point of the trunk on the dorsal side. The *second dorsal fin* is directly caudal to the first dorsal fin. The cranial dorsal fin is supported by *hard fin rays* (spines) but the second dorsal fin is *soft* (soft fin rays) and spineless.

The caudal end of the trunk and the beginning of the tail are marked by the external orifices—anal and urogenital (urinary and genital) openings are separate in the male (see chap. 9). Just caudal to these openings is a ventral *anal fin.*

TAIL

The *tail* of the perch tapers from the trunk but ends rather abruptly as a laterally compressed "paddle." The anal fin just mentioned is a soft-rayed fin like the second dorsal fin. The fleshy end of the tail is surrounded by a symmetrical *caudal fin.*

The type of caudal fin on the perch (and other bony fishes) is called *homocercal* (ho-mo-SER-kal). Sharks have a *heterocercal* (het-er-o-SER-kal) tail fin and the lamprey has a *diphycercal* (dif-i-SER-kal) tail (fig. 2.3). The type of caudal fin may be correlated with the presence (homocercal) or absence (heterocercal) of a *swim bladder* (BLAD-der) and the method of locomotion. Sharks, for example do not have a swim bladder and the density of the shark's body is greater than water so the shark tends to sink. To counteract this the shark must be constantly swimming. Movements of the shark's heterocercal tail in swimming tends to drive the head down, a consequence as bad as the loss of a swim bladder. To counteract the downward force of the tail, the shark's paired pectoral fins are placed ventrally and at an angle that deflects the head upward.

The perch has a swim bladder so the caudal fin-pectoral fin relationship is unnecessary. The fins of the perch may be used as rudders for more diversified movements and as brakes for sudden stops. Sharks cannot brake their forward movement. In order to avoid obstacles the shark must turn sharply.

Some bony fishes with homocercal tails have secondarily "lost" the swim bladder (or lungs) and are consequently relegated to a bottom dwelling life. The occasional excursions into more shallow levels require considerable effort on the part of these buoyantless forms.

HISTOLOGY OF THE SKIN

The skin consists of two parts (fig. 2.4): an outer *epidermis* (ep-i-DER-mis) and an inner *dermis* (DER-mis).

The dermis is derived from mesenchyme (MES-en-kim) of the mesoderm (MES-o-derm) which also forms bone, muscle, and blood vessels as well as connective tissue, and is the predominant tissue of the dermis.

The epidermis is the most prominent derivative of ectoderm (EK-to-derm) and in fishes it is composed entirely of living cells. The epidermal cells nearest the dermis are square (cuboidal) or tall (columnar). These cells represent the generative cell layer of the epidermis or the *stratum germinativum* (STRA-tum jer-mi-na-T*I*V-um). New cells form by asexual cell division and migrate toward the surface. As they migrate, the cells become progressively flatter (squamous) and at the surface of the epidermis the flattened cells are sloughed off and replaced by new cells from the deeper epidermal layers.

In several epidermal cells the cytoplasm of the cell undergoes a decomposition to form mucus. To some small extent this mucus may help to restrict the passage of water through the skin as well as reduce profile drag during forward movement.

The dermis is composed of a dense fibrous connective tissue containing bony *scales* and is attached to the underlying muscles by a *subcutis* (sub-KU-tis). The subcutis is a loose network of connective tissue. Pigment (color) cells are present in both the epidermis and subcutis.

The scale of the perch is a thin, flexible plate composed of two parts. The superficial portion (fig. 2.4) of the scale is the so-called *bony layer.* Presumably, the bony portion of the scale has most of the inorganic material, but there are exceptions. The deeper or inner portion of the scale is called the *fibrous layer* and it is composed of several layers of fibrous connective tissue. Each connective tissue layer is arranged at right angles to each of the layers adjacent to it. The result is a crisscross of fibers that make the scale strong but flexible. In a few species of fish the fibrous layer may be imbedded with some calcium salts, but this is not usually the case in the perch.

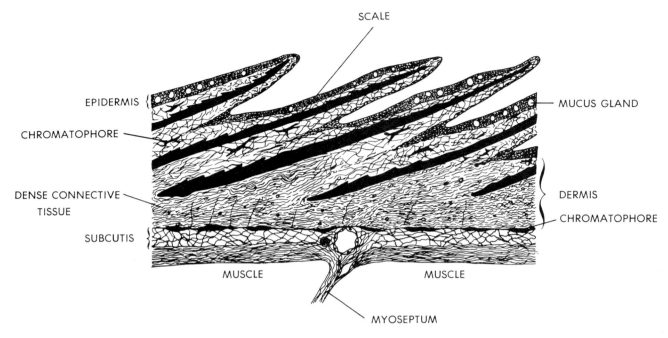

Figure 2.4 Semidiagrammatic cross section of the skin.

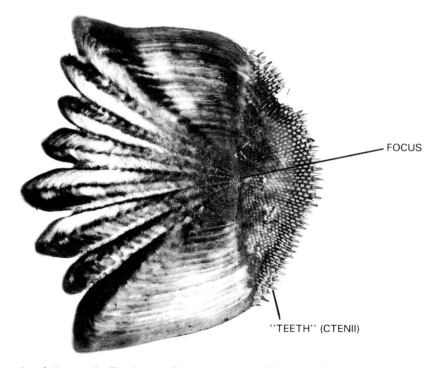

Figure 2.5 Ctenoid scale of the perch. Teeth (ctenii) are on the caudal exposed portion of the scale. The cranial embedded portion of the scale has grooves or radii which radiate from a central focus. [Photograph by permission of General Biological Supply House, Chicago.]

The scale of the perch is called *ctenoid* (TEN-oyd) (fig. 2.5) because the exposed, caudal portion bears small spiny structures called *ctenii* (TEN-i). In most instances ctenoid scales are found on those fishes with stiff spines in their fins. Soft-rayed fishes have scales that usually lack ctenii and this type of scale is termed *cycloid* (SI-kloyd) (fig. 2.6). The *placoid* (PLAK-oyd) scale is found in the cartilaginous fishes and has a backward-pointing spine. Fish scales are useful in age and growth studies of fishes.

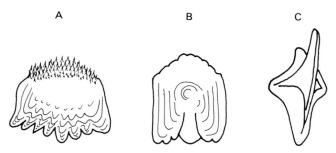

Figure 2.6 A comparison of (A) ctenoid, (B) cycloid, and (C) placoid scales.

SUGGESTED READINGS

Bhatia, D. 1931. On the production of annual zones in the scales of the rainbow trout (*Salmo irideus*). *I. J. Exp. Zool.* 59:45–60.

Brown, G. A. and S. R. Wellings. 1969. Collagen formation and calcification in teleost scales. *Z. Zellforsch.* 93:571–582.

Geraudie, J. and F. J. Meunier. 1982. Comparative fine structure of the osteichthyan dermotrichia. *Anat. Rec.* 202(3):325–328.

Gray, J. and S. B. Setna. 1931. The growth of fish. *J. Exp. Biol.* 8:55–62.

Green, E. H. and R. W. Tower. 1902. The organic constituents of the scales of fish. *Bull. U.S. Fish Comm.* 21:97.

Hawkes, J. W. 1974. The structure of fish skin. I. General organization. *Cell Tiss. Res.* 149:159–172.

———. 1974. The structure of fish skin. II. The chromatophore unit. *Cell Tiss. Res.* 149:159–172.

Hendrickson, R. S. and A. G. Matoltsy. 1967. The fine structure of teleost epidermis. 1. Introduction and filament-containing cells. *J. Ultrastruct. Res.* 21:194–212.

Meunier, F. J. and J. Geraudie. 1980. The plywood-like arrangement of structures of the skin and scales of lower vertebrates. *Annee Biol.* 19(1):1–18.

Mittal, A. K. and M. Whitear. 1979. Keratinization of fish skin with special reference to the catfish (*Bagarius bagarius*). *Cell Tiss. Res.* 202(2):213–230.

Nashimoto, K. 1981. The swimming speed of fish in relation to frequency of tail beating and body type. *Bull. Jpn. Soc. Sci. Fish.* 46(6):675–680.

Neave, F. 1936a. Origin of the teleost scale-pattern and the development of the teleost scale. *Nature* (London) 137:1034–1035.

———. 1936b. The development of scales of *Salmo*. *Trans. Roy. Soc. Can.* Sect. V. 30:55–72.

———. 1940. On the histology and regeneration of the teleost scale. *Quart. J. Micr. Sci.* 81:541–568.

Oosten, J. Van. 1957. Skin and scales. *The physiology of fishes* (M. E. Brown, ed.). Vol. 1, pp. 207–244. New York: Academic Press, Inc.

Shetyakova, L. F. 1966. Some biological and topographic characteristics of fish scales. *Gidrobiolzh* 2:60–67.

Thomson, K. S. and D. E. Simanek. 1977. Body form and locomotion in sharks. *Amer. Zool.* 17(2):343–354.

Wallen, O. 1957. On the growth structure and developmental physiology of the scale of fishes. *Drottningholm Rep.* 38:385–447.

Webb, P. W. 1984. Body form, locomotion and foraging in aquatic vertebrates. *Amer. Zool.* 24:107–120.

Zaets, V. A., A. P. Koval, and T. A. Kalyuzhnaya. 1981. Interaction of skeleton-forming and slime-forming function of the skin. *Vestn. Zool. O.* 2:52–56.

Zylberberg, L. and G. Nicolas. 1983. Ultrastructure of scales in a teleost (*Carassius auratus*) after use of rapid freeze fixation and freeze substitution. *Cell Tiss. Res.* 223(2):349–367.

Chapter 3
The Skeletal System

Prepared skeletons of the perch will be necessary for this study. The bones of the prepared skeletons are delicate and easily broken. Be especially careful in handling these skeletons. Do not use pencils or pens for pointers because these will permanently mark bone. Use a dissecting needle or metal probe when pointing out bones to your neighbor or to the instructor. If damage occurs, please alert your instructor immediately so that the item may be repaired before parts are lost.

The skeleton of the perch is composed of bone and consists of *axial* (AK-se-al) and *appendicular* (ap-en-DIK-u-lar) regions.

The bone of vertebrates forms from two sources:

1. A *dermal* (DER-mal) component originates directly from mesenchyme in the dermis of the skin. Dermal bone forms in the dermis of the skin in a manner somewhat similar to the formation of scales; that is, the bone is deposited directly from the embryonic cells of the dermis and is not preceded by a cartilage structure. This is the bone of the most primitive fishes (Ostracoderms and Placoderms; see fig. 1.1).
2. An *endochondral* (en-do-KON-dral) component is derived from deeper tissues and generally forms first as cartilage to be later replaced by bone.

THE AXIAL SKELETON

The axial skeleton includes the bones of the *skull*, *vertebrae* and *ribs*.

The Skull

The skull of the perch is actually a double skull. The outer skull is an armor of *dermal bone*, the *dermatocranium* (der-ma-to-KRA-ne-um). The inner skull is composed of bone formed deep in the body (subdermal) and the individual skeletal structures are usually preceded in development by a corresponding cartilage structure; hence, the name *chondrocranium* (kon-dro-KRA-ne-um) or *neurocranium* (nu-ro-KRA-ne-um). The *splanchnocranium* (SPLANK-no-kra-ne-um) is a portion of the head skeleton that supports the gill arches and their derivatives.

I. The Dermatocranium
   A. The Skull Roof (figs. 3.1 and 3.2)
      1. *Nasal* (NA-sal) bones are small, paired, rostral bones bordering the nasal capsule. These small bones are widely separated in the midline. Prepared skulls should be carefully cleaned of dried connective tissue in the region of the nasal bones. In most preparations, these bones are completely obscured by dried tissue.
      2. *Frontal* (FRUN-tal) bones are also paired dermal bones. These are the largest bones of the dermal skull roof. The frontals make up most of the head shield including the dorsal border of the orbit. Some authors call these bones the parietals and the following set of bones are then called postparietals.
      3. *Parietal* (pah-RI-e-tal) bones are very small paired bones just caudal to the frontals. These bones are separated in the midline by the dorsal spine of the supraoccipital.
      4. *Scale bones,* one on each side, form an angle that unites the epiotic (see II., number 7) with one arm of the Y-shaped posttemporal.
      5. *Posttemporals* (pos-TEM-po-ral) form the dorso-caudal corner of the dermatocranium.
   B. Circumorbital Bones (ser-kum-OR-bi-tal) (fig. 3.1)
      1. The *lacrimal* (LAK-ri-mal) is the most rostral and the largest of the circumorbital bones. There are no supraorbital bones in the perch; instead, the frontal bone is the dorsal border of the orbit (fig. 3.1). The remaining circumorbital bones are all suborbital.

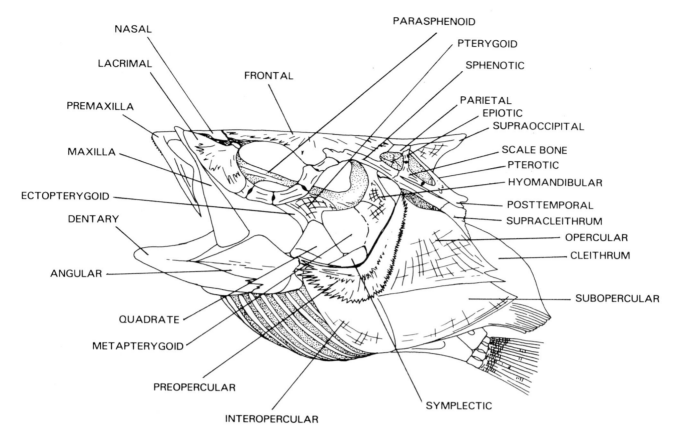

Figure 3.1 Lateral view of the skull.

2. *Suborbital* (sub-OR-bi-tal) bones are a series of four small dermal bones extending from the lacrimal to the frontal and sphenotic bones.
3. The *sphenotic* (SFE-not-ik) continues caudally from the suborbitals and at the lateral edge of the frontals. The sphenotic is a long slender bone that abuts the dorsal scale bone (see A., number 4 above) and a ventral pterotic.
4. The *pterotic* (TER-ot-ik) runs parallel to the bar formed by the combined scale bone and posttemporal (see A., numbers 4 and 5 above). Caudally, the pterotic contacts the supracleithrum of the pectoral girdle.

C. Jaws and Palate (fig. 3.3)
 1. *Premaxillae* (pre-mak-SIL-ah) are the most rostral bones of the upper jaw. The two bones are joined rostrally in a *symphysis* (SIM-fi-sis). The ventral (palatal) surface is lined with teeth. The slender caudal tips of the premaxillae are embedded in connective tissue that binds them loosely to the caudal ventral margin of the maxillae. There are two rostral dorsal processes of the premaxillae. The most rostral of these processes is the *ascending process* that articulates with the dorsal process of the maxillae and the rostral tip of the nasal. Caudal and lateral to the ascending process is the *articular process* (ar-TIK-u-lar) with a rounded articular head that fits in the concave socket of the *premaxillary condyle* (KON-dil) of the maxilla. A mid-dorsal lateral process of the premaxilla helps to guide the bone as it moves on the medial surface of the maxilla.
 2. The *maxillae* (mak-SIL-ah) are paired bones lying dorsal to the premaxillae and articulating with them. Caudally, the medial surface of the maxilla has a tendinous articulation with the lateral tip of

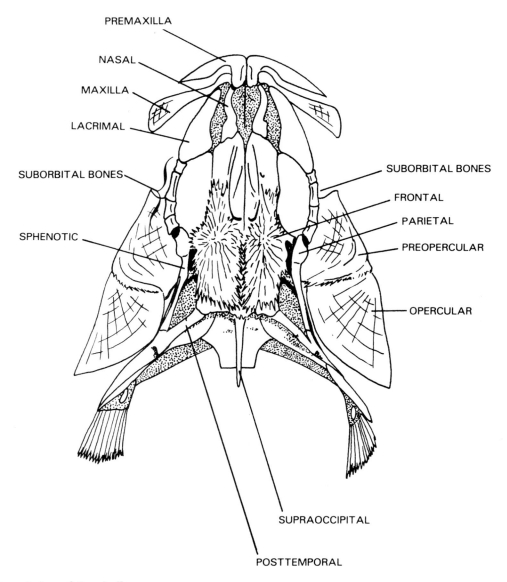

Figure 3.2 Dorsal view of the skull.

the dorsal process of the dentary. Rostrally, each maxilla has three articular processes. A prominent *dorsal process* has an ethmoid-vomer condyle that articulates with these two medial bones. A forward protruding *rostral process* bears a "saddle-shaped" articulation for the palatine bone. Ventral to the ethmoid-vomer condyle is a *medial process* with a prominent *premaxillary condyle*. The maxillae do not bear teeth.

3. The *angular* (ANG-gu-lar) bone is a small bone at the caudal lower angle of the lower jaw. This bone and the dentary (below) are the dermal bones surrounding the endochondral articular bone.

4. The *dentary* (DEN-tar-*e*) is the tooth-bearing bone of the lower jaw. Each dentary is roughly V-shaped with the point of the V directed rostrally. The large articular fits inside the ventral limb of the V.

5. The *articular* (ar-TIK-*u*-lar) is the bone of the lower jaw with an articular surface for the reception of the quadrate. This bone represents the caudal (or proximal) portion of the primary lower jaw, *Meckel's cartilage* (MEK-elz). The

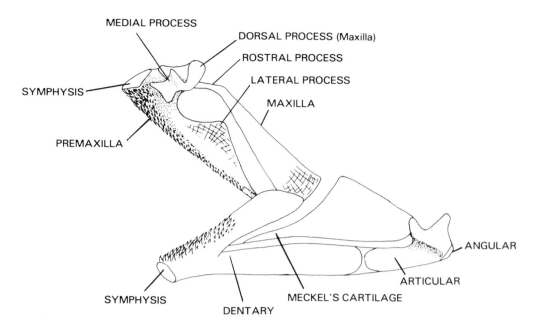

Figure 3.3 Medial view of the right jaws.

articular is an endochondral bone forming the mandibular (lower jaw) portion of the jaw joint. This juncture between the quadrate and articular is the jaw articulation of all vertebrates except mammals.

6. The *vomer* (V*O*-mer) (fig. 3.4) is a small, median, unpaired dermal bone just caudal to the maxillae. This is the most rostral bone of the palate of the perch. Caudally, the vomer articulates with the palatines. The vomer and both palatine bones have teeth.

D. The Opercular Series (fig. 3.1)
1. The *opercular* (*o*-PER-kyoo-ler) is a large rhomboid-shaped bone just caudal to the vertical limb of the preopercular "L".
2. The *interopercular* forms the rostral ventral border of the operculum.
3. The *subopercular* forms the caudal ventral border of the operculum.

II. The Neurocranium
The Cranium
1. The *parasphenoid* (par-a-SF*E*-n*o*yd) (fig. 3.4) is a long, slender unpaired bone extending from the vomer to the endocranium (basioccipital) between the two pterygoids.

2. The *basioccipital* (ba-se-ok-SIP-i-tal) is the true floor of the braincase. The slender parasphenoid separates the basioccipital and the toothed palate (vomer and palatines). Laterally, the two exoccipitals join the basioccipital and form the walls of the cranial box.

3. The *exoccipitals* (ek-sok-SIP-i-tal) in caudal view (fig. 3.5) resemble the transverse processes of vertebrae. The caudal face of each exoccipital is expanded to receive the expanded cranial face of the prezygapophysis of the first vertebra (see fig. 3.9).

4. The *supraoccipital* (syoo-pra-ok-SIP-i-tal) serves as the roof of the braincase and provides a large spine for the attachment of trunk muscles.

5. Lateral *ethmoid* (ETH-m*o*yd) bones (not illustrated) extend dorsally from the rostral end of the parasphenoid and support the rostral end of the parasphenoid, the rostral end of the frontal bones, and the caudal end of the lacrimals.

6. The *mesethmoid* (MES-eth-m*o*yd) (not illustrated) is an unpaired median bone just rostral to the lateral ethmoids. This bone is based on the vomer and contacts the caudal end of the nasals and the rostral end of the frontals.

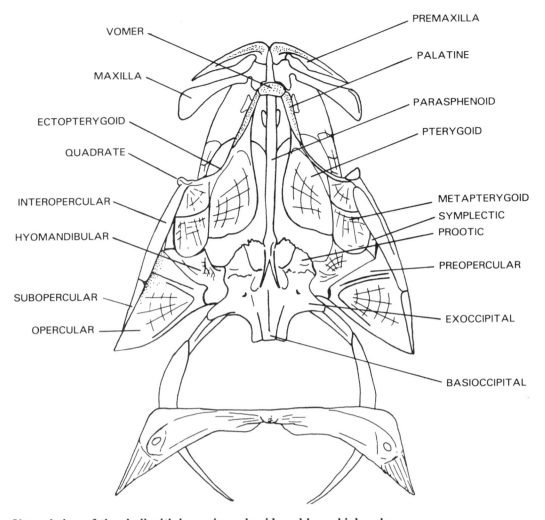

Figure 3.4 Ventral view of the skull with lower jaws, hyoid, and branchial arches removed.

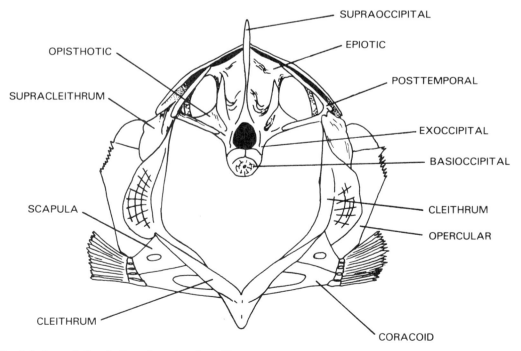

Figure 3.5 Caudal view of the skull and pectoral girdle.

7. An *epiotic* (ep-*e*-*O*T-ik) bone abuts each parietal bone. Each epiotic bone together with a scale bone and a posttemporal (see p. 14), make up a system of caudal struts that arch laterally from the parietal and supraoccipital. The epiotic bone may be seen on skulls in which the occipital region is fairly well cleaned. Both the epiotic and opisthotic bones are contacted by the posttemporal (see fig. 3.5), which is a **V**-shaped bone with the point of the **V** directed caudally and the two arms arranged so one arm is dorsal and the other is ventral. The dorsal arm contacts the epiotic and the ventral arm contacts the opisthotic.

8. The *opisthotic* (*o*-pis-THOT-ik) is the most caudal portion of the *otic* (*o*-tik) capsule. This bone and the preceding (epiotic) are probably the only bones of this series that you will be able to see.

9. The *prootic* (pr*o*-*O*t-ik) is at the ventro-rostral corner of the cranial box. The rostral dorsal part and the caudal dorsal part of the hyomandibular contacts this bone, and the caudal dorsal part of the hyomandibular contacts the epiotic.

10. The *sphenotic* (sfe-N*O*T-ik) (fig. 3.1) is dorsal to the prootic and rostral to the epiotic. These bones (epiotic, opisthotic, prootic, and sphenotic) contain the semicircular canals.

III. The Splanchnocranium
   A. The Suspensorium (This structure supports the lower jaw.)
      1. The *hyomandibular* (hi-*o*-man-DIB-*u*-lar) is located between the quadrate and the cranium (see number 3 that follows). This bone represents a partial remnant of the primitive first branchial arch which, together with a ventral portion (ceratohyal, epihyal, and interhyal), forms the *hyoid* (H*I*-*o*yd) arch. In tetrapods the hyomandibular is associated with the ear to conduct sound vibrations from the *tympanum* (tim-PA-num) to the inner ear. In these vertebrates the bone is known as the stapes (ST*A*-pez). Ventrally, the hyomandibular of the perch contacts two bones, a rostral metapterygoid (see number 7 that follows) and a caudal symplectic. Dorsally, the hyomandibular has a small process that articulates in a socket of the *operculum* (*o*-PER-k*u*-lem).

      2. The *symplectic* (sim-PLEK-tic) is braced by the hyomandibular and aids the metapterygoid in the support of the quadrate. The symplectic is really a small splinter of the hyomandibular wedged between the preoperculum and metapterygoid.

      3. The *quadrate* (KWOD-r*at*) contains the articular surface of the upper jaw that fits in the lower jaw. Although it is usually noted that the quadrate is suspended by the hyomandibular (and symplectic), it is easily seen that the quadrate also gains considerable support from the ventral limb of the preoperculum.

      4. The *palatines* (PAL-a-t*i*ns) are narrow bones extending laterally and caudally from either side of the vomer. Caudally, the palatines join the ectopterygoids to form a solid brace between the rostral palate (the vomer) and the quadrate bone.

      5. The *ectopterygoid* (EK-t*o*-ter-i-g*o*yd) bones are similar in appearance to the palatines, except the ectopterygoids do not bear teeth. Each ectopterygoid bone articulates with a palatine rostrally, the quadrate caudally, and with the pterygoid medially.

      6. The *pterygoids* (TER-i-g*o*yd) are thin bones forming most of the roof of the mouth and the floor of the orbit of the eye; also termed "entopterygoid."

      7. The *metapterygoid* (MET-ah-ter-i-g*o*yd) continues with the caudal edge of the pterygoid and helps to support the caudal border of the quadrate.

      8. The *preoperculum* (pre-*o*-PER-k*u*-lem) is an **L**-shaped bone with a scalloped caudal border. This bone forms a base for the other members of the opercular series.

B. Hyoid Apparatus and Gill Arches (figs. 3.6 and 3.7). Part of the hyoid apparatus has been described as a part of the jaw suspensorium (see III., A. above). The remainder of the hyoid serves as a base for the gills. Most of this apparatus is of endochondral bone but a few exceptions are noted.
   1. The *ceratohyals* (ser-a-to-H*I*-al) are the main body of the hyoid apparatus. These bones are separated in the rostral midline by the hypohyal. Caudally, they have an attached epihyal and ventrally, a series of branchiostegal rays are based on the ceratohyal.
   2. The *epihyal* (ep-i-H*I*-al) is a functional continuation of the ceratohyal. One branchiostegal ray is based on the epihyal and another is attached at the joint between the epihyal and ceratohyal.
   3. An *interhyal* (in-ter-H*I*-al) suspends the entire apparatus from the hyomandibular (see III., A., number 1).

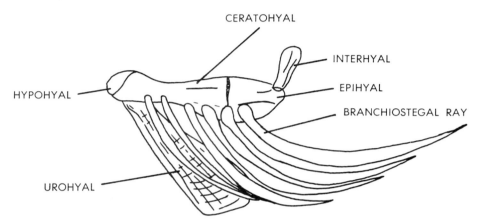

Figure 3.6   Lateral view of the hyoid apparatus.

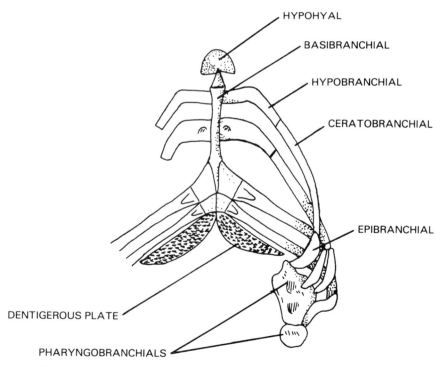

Figure 3.7   Dorsal view of the branchial apparatus.

4. Seven *branchiostegal rays* (brank-*e*-*o*-ST*E*-gal) extend ventrally and caudally from each side of the hyoid arch. These bones are dermal structures.
5. The *hypohyal* (h*i*-po-H*I*-al) serves as a link between the right and left halves of the hyoid apparatus, and as a base for a dermal bone (the urohyal) that forms in the connective tissues between the long throat muscles.
6. The *urohyal* (*u*-ro-H*I*-al) is a dermal, not endochondral, bone, but it is intimately associated with the hyoid apparatus.
7. The *gill apparatus* is a complicated series of four bony arches that support the gills. Each arch is based ventrally on a *basibranchial* (b*a*-se-BRANK-*e*-al) and consists of (from ventral to dorsal) a *hypobranchial,* a *ceratobranchial,* and an *epibranchial.* Dorsally, each epibranchial is attached to a fused *pharyngobranchial* complex. The first three pharyngobranchials are fused together and the last pharyngobranchial articulates with the rostral mass. Dermal, tooth-bearing bones are fused to the ventral surface of the pharyngobranchials. At the caudal end of the last gill arch, a dermal tooth-bearing plate is attached to each terminal hypobranchial.

THE VERTEBRAL COLUMN

The vertebral (fig. 3.8) column of the perch is composed of a series of endochondral skeletal units termed *vertebrae* (VER-te-br*a*). Each of the vertebrae have certain common features. The large "spool-shaped" central portion is the *centrum* (SEN-trum) or body (fig. 3.9). Notice that both the cranial and caudal faces of the centrum are funneled toward the center. This double concave centrum is termed *amphicoelous* (am-fe-S*E*-lus). At the middle of the centrum is a *notochordal canal* (no-to-K*O*RD-al) for the passage of the notochord. Fishes retain the notochord throughout life. Dorsal to the centrum is an arch, surmounted by a spine. The arch protects the spinal cord, which passes though it, and is therefore termed the *neural arch* (NU-ral). The spine is called the *neural spine.*

The remaining features are characteristics of the vertebrae of particular regions of the column. The two major subdivisions of the fish vertebral column are the trunk and the tail. Although there is no neck in the fish, the first two trunk vertebrae are modified and appear quite different

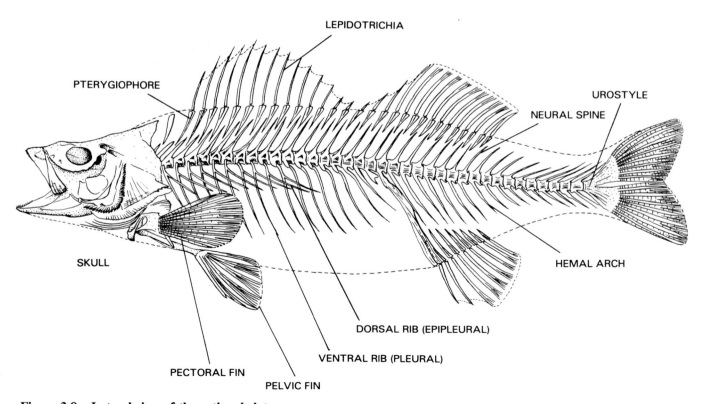

Figure 3.8 Lateral view of the entire skeleton.

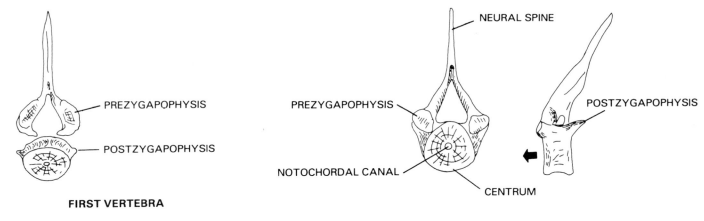
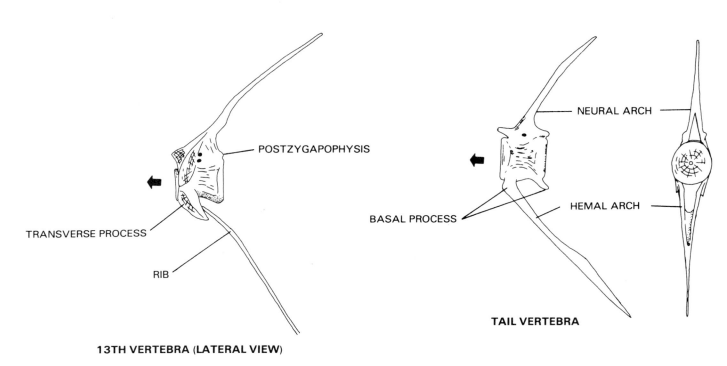

Figure 3.9 Representative vertebrae. Arrows indicate cranial.

from the remaining vertebrae. Some authors have referred to these vertebrae as the *atlas* (AT-las) and *axis* (AX-sis), but the terms seem to be presumptive. The "atlas" does not support the fish skull and the "axis" does not serve as a pivot for the rotation of the head as they do in terrestrial vertebrates.

The first vertebra of the perch is in two pieces. The dorsal piece is composed of a neural arch and neural spine and a pair of expanded *cranial articular processes* or *prezygapophyses* (pre-zi-ga-POF-i-ses) at each side of the base of the neural arch. The ventral piece consists of the centrum with two small *caudal articulating processes* or *postzygapophyses*—one on each side of the dorsal edge of the caudal face of the centrum. The zygapophyses are not interlocked as in tetrapods, but instead, serve only as surfaces of contact that probably restrict movement.

The second vertebra of the perch is a single unit, but aside from this difference the two vertebrae are very similar. The second vertebra, like the first, has a large prezygapophysis. Both lack transverse process and ribs and they are both the same size. The zygapophyses of the remaining vertebrae are all small structures and in many instances fail to make contact. The transverse processes look like a tiny leaf extending laterally from the cranial ventral corner of the connective tissues. There are no joints or articular surfaces involved in rib attachments. The number of vertebrae vary somewhat among individuals, but there appears to be an even more important variation in the total number of vertebrae between widely separated populations. A population of perch in Maine had 41 vertebrae with rare individuals having as few as 40 or as many as 42. In Michigan, a perch population had 40 vertebrae with individual variation from 39 to 41. (Bailey and Gosline, 1955). The ribs of the perch are of two types (fig. 3.8). A *pleural* (PLOO-ral) or *ventral* rib develops in the *myosepta* (mi-o-SEP-ta) just lateral to the pleuroperitoneal lining (see chap. 6). The *epipleural* (ep-ih-PLOO-ral) or *dorsal* rib develops in the horizontal septum and attaches to the caudal surface of the pleural ribs by ligaments. Since the epipleural rib is between the epaxial and hypaxial muscle masses (see chap. 4), it is sometimes called an *intermuscular bone* (in-ter-MUS-kyoo-lar). The epipleural ribs are only attached to the cranial group of the pleural ribs; the last seven or eight pairs of pleural ribs do not have epipleurals attached to them. Some teleosts have an additional set of ribs based on the upper portion of the centrum or on the base of the neural arch. In the tail the ribs and traverse processes are replaced by a *hemal arch* (HEM-al) similar to the neural arch (fig. 3.9). The hemal arch surrounds blood vessels rather than nerves. Hemal arches of the caudal vertebrae are equipped with processes similar to the pre- and postzygapophyses but they do not contact one another. These processes may be referred to as basal processes of the hemal arch.

THE APPENDICULAR SKELETON AND FINS (fig. 3.8)

I. The Medial Fins
   A. Dorsal
      1. The *first dorsal fin* has dermal structures for fin support as do all the fins. These dermal structures act as fin rays and are thought by some authors to be evolved from scales (see chap. 2). In bony fishes the fin rays are of two types. Some fin rays are *ossified* (OS-ih-fid) providing a stiff bony support for the fin and are known as *lepidotrichia* (lep-ih-do-TRIK-e-ah) or "spines." Others are not ossified but are flexible, typically segmented, and often branched. These are called *ceratotrichia* (ser-ah-to-TRIK-e-ah). The cranial dorsal fin of the perch consists of spines only. The base of each spine or soft ray is supported by a bony structure known as a *pterygiophore* (te-RIJ-e-o-for) (one pterygiophore for each fin ray). The inner tip of the pterygiophore lies in the connective tissue between the neural spines of the trunk vertebrae. With this arrangement, the musculature contorting the fish's body also bends the fin without collapsing it against the body.
      2. The *second dorsal fin* is actually a continuation of the first dorsal fin. The pterygiophores are continuous from first to second dorsal fin as are the fin rays. The separation between the two fins is marked by a reduction in the length of the spines between the fins. Most of the second dorsal fin rays are soft but the first two are spiny. Even the soft rays often have an ossified proximal segment (the segment based on the pterygiophore). The distal, unossified portion of the soft ray is segmented and often branched.
   B. Ventral

The *anal fin* is located just caudal to the anal and urogenital openings. The first two fin rays of the anal fin are ossified but the remaining ones are typical soft rays. The cranial pterygiophores of the anal fin are usually fused into a single large base for the first two or three fin rays. The internal "spine" of the enlarged pterygiophore fuses with the last one or two pairs of ribs on each side. The first true hemal arch is just behind the "spine" of the enlarged pterygiophore. The remaining pterygiophores are shorter and the inner tips of these pterygiophores barely reach the tips of hemal spines.

   C. Caudal

The *caudal fin* is composed entirely of soft rays. The caudal fin rays are based on modified parts of the caudal vertebrae, not on pterygiophores. All of the major parts of the vertebrae seem to be involved in the formation of the fin base (fig. 3.10). In general, the neural spines of the vertebrae are compressed into two or three plates called *epurals* (ep-UR-als). The last caudal vertebra is modified into a dorso, caudally directed plate, the *urostyle* (U-ro-stile). Below the urostyle are six or seven modified hemal arches, termed *hypurals* (HIP-ur-als). A few unmodified neural and hemal arches and spines may assist the modified caudal units.

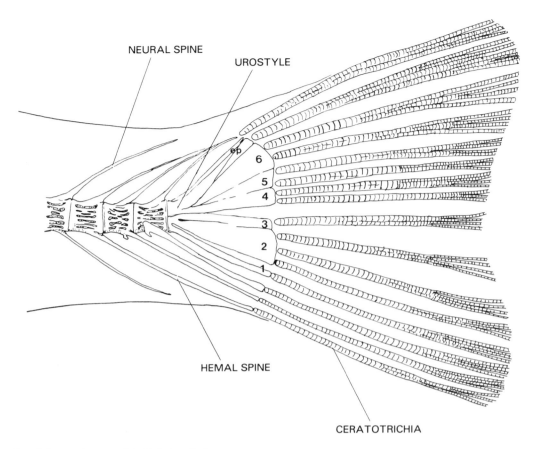

Figure 3.10 Caudal vertebrae and tail, lateral view.

II. The Appendicular Skeleton
   A. The Pectoral Girdle and Fins (fig. 3.11)
      1. The *cleithrum* (KL*I*-thrum) is the major bone of the pectoral girdle. The two cleithra are joined to one another in the ventral midline. Dorsally, each cleithrum is linked by a supracleithrum to the skull.
      2. The *supracleithrum* is the link between the girdle and the skull. The caudal end of the supracleithrum joins the cleithrum, and the rostral end contacts the posttemporal bone (see p. 14, number 5). The posttemporal bone is occasionally considered a bone of the pectoral girdle rather than of the skull. In this case, the otic capsule serves as the skull bone contacting the girdle (see *epiotic,* page 19).
      3. The *scapula* (SKAP-*u*-la) and the coracoid form the distal, second tier of the pectoral girdle. The scapula is the dorsal of the two bones based on the cleithrum. Distally, the scapula serves as a base for half of the fin radials.
      4. The *coracoid* (K*O*R-a-koyd) is ventral to the scapula and, like the scapula, the coracoid is also based on the cleithrum. The perch coracoid is a little larger than the scapula. In tetrapods that have lost the cleithrum, the scapula becomes the more important of the remaining girdle bones.
      5. The *postcleithrum* is a fan-shaped bone held to the medial surface of the pectoral girdle by connective tissue. The flattened blade of the "fan" is just medial to the

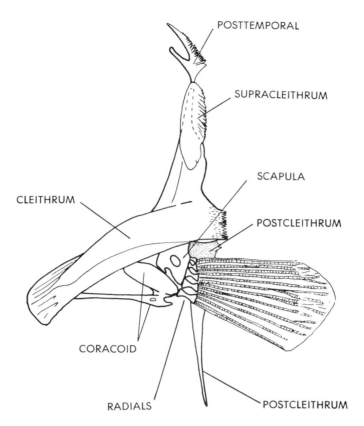

Figure 3.11 Lateral view of left pectoral girdle and limb.

cleithrum and scapula. The long-pointed spine of the "fan handle" extends ventrally, towards the pelvic girdle.

6. *Radials* (R*A*-de-als) are bones of the fin with two joints. One joint allows movement between the radial and the scapula or coracoid, and the other joint allows movement between the radial and the fin rays. There may be some variation in the number of radials but there are usually only four in the perch pectoral fin. Soft rays support the remainder of the fin.

B. Pelvic Girdle and Fins (fig. 3.12) The paired *pelvic plates* or *basipterygia* (bas-*e*-te-RIJ-*e*-a) make up the entire pelvic girdle of the perch. The two plates may be fused to each other in the midline. The rostral tips of the pelvic plates extend forward between and dorsal to the joined cleithra. Some authors have attempted

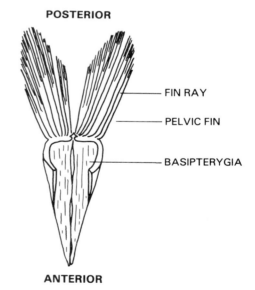

Figure 3.12 Pelvic girdle and fins of the perch. Ventral view.

to homologize these bones with the *pubic* (PUbik) bones of tetrapods. There are no radials at the base of the pelvic fin. Without radials, the fin rays are based directly on the caudal ends of the pelvic plates. The fin rays are of the typical soft-rayed nonossified type except for the first (most medial), which is spiny.

SUGGESTED READINGS

Bailey, R. M. and W. A. Gosline. 1955. Variation and systematic significance of vertebral counts in the American fishes of the family Percidae. *Mus. Zool. Univ. Mich., Misc. Publ.* 93:1–44.

Bond, C. E. and T. Uyeno. 1981. Remarkable changes in the vertebrae of Perciform fish *Scombrolabrax* with notes on its anatomy and systematics. *Gyoruigaku Zasshi Jap. J. Ichthyol.* 28(3):259–262. Tokyo, Nihon Gyorui Gakkai.

Hebrank, M. R. 1982. Mechanical properties of fish backbones in lateral bending and in tension. *Jour. Biomechanics* 15(2):85–90.

Lee, R. F., C. F. Phleger, and M.H. Horn. 1975. Composition of oil in fish bones: Possible function in neutral buoyancy. *Comp. Biochem. Physiol.* 50(B):13–16.

Lehman, J. P. 1979. The anatomy of the endocranium of the Actinopterygians. *Bull. Mus. Nat. Hist. Nat. Sect. C. Sci. Terre Paleontol. Geol. Mineral* 1(4):367–370.

Nelson, E. M. 1963. A preparation of a standard teleost study skull. *Turtox News* 41(2):72–74.

Osse, J. W. M. 1969. Functional morphology of the head of the perch (*Perca fluviatilis* L.). An electromyographic study. *Netherlands Jour. of Zoology* 19(3):289–392.

Schultze, H. P. and Arratia, G. 1989. The composition of the caudal skeleton of teleosts (Actinopterygii, Osteichthyes). *Zool. Jour. Linn. Soc.* 97(3):189–231.

Smithson, T. R. and K. S. Thomson. 1982. The hyomandibular of *Eusthenopteron foordi*, Pisces, Crossopterygii and the early evolution of the tetrapod stapes. *Zool. Jour. Linn. Soc.* 74(1):93–103.

Chapter 4
The Muscular System

PREPARATION OF THE SPECIMEN FOR OBSERVATION OF THE TRUNK MUSCULATURE

The skin around the dorsal fin or on the belly of the perch is more easily separated from the underlying muscles than is the skin on the sides of the body or over the skull. Pinch the skin near the dorsal fin with forceps and cut an opening with the tip of a pair of scissors. Be careful not to cut through the underlying muscles. Separate the skin from the underlying muscles by inserting a blunt probe through the opening you have cut through the skin and move the probe about under the skin. Next grasp the edge of the cut skin and tear the skin off. If the skin adheres to the muscle, use a blunt probe to separate the connective tissue. Continue to tear the skin off in pieces until all of it is cleaned away from the trunk (leave the head skin intact at this time). With the skin removed it is necessary to keep the specimen moist at all times and especially during storage to prevent drying (use water). Cover the specimen with wet paper toweling during breaks or whenever you are not actively dissecting or observing the animal.

The muscles of the perch are not as complex as the muscles of land vertebrates; and yet considering relative body sizes, the fish probably has twice the muscle mass of most tetrapods. The reasons for this contrast in the amounts of muscle are evident if we consider two important differences between tetrapods and fishes. First, the fish moves by lateral undulations of the trunk not limb movements; and, second, the fish muscles are not isolated units producing unique movements of a specific structure as in tetrapods. The contractions of the muscle fibers within a fish *myotome* (M*I*-o-tom) exert their action indirectly upon the vertebral column (a myotome is the portion of a somite that develops into voluntary muscle). Thus, the relative inflexibility of the fish trunk and the indirectness (hence, inefficiency) of the fish muscles demand more muscle mass to perform less action than in tetrapods.

BODY MUSCULATURE

Each perch myotome looks like a large "W" tilted on its side. The two bottom points of the "W" are directed caudally and the center upper point is cranial (see fig. 4.1). Each of the three points of the "W" is related to the corresponding point of the preceding or succeeding myotome like one of a stack of paper cups is related to a preceding or succeeding cup. That is, the myotome does not extend directly from the midline to the surface, but angles, cranially on the dorsal and ventral points, and caudally on the middle point. Thus, if a horizontal section is cut so the myotome is divided at any of these three points, we would find the muscle segment shaped as in figure 4.2. Each muscle fiber is arranged parallel to the midline and extends from one myosepta to the next. The simultaneous contraction of all the fibers at this level tends to move the two myosepta toward one another. This actually may happen to a slight extent but several factors work against it. If the myosepta were to move to a position, parallel to its resting position, then any point on the myosepta may have a movement vector perpendicular to the surface of the myosepta. We may indicate this as a vector of the force of the contracting muscle fiber that acts with a second vector, 90 degrees from the first, to move the myosepta. The vectors on either end of the myotome that are perpendicular to their respective myosepta surfaces will cancel each other out because they are equal in strength and directly opposed. The effective vectors are parallel to the myosepta surfaces and tend to torque the myotome (see fig. 4.3). If a caudal medial myotome segment is opposed to a cranial dorsal (or ventral) myotome segment, the opposing torques will bend the midline of the body at a point midway between the two myotomes. A second consideration of myotome contraction concerns shape changes. Each muscle cell or fiber contracts one-half its length. If a single cross-sectional area of muscle fibers was to contract simultaneously, the cranial-caudal distance is reduced by one-half. Since the volume does not change, the myotome must expand laterally. The slanting arrangement of the myotomes presents less resistance to the lateral expansion of the muscle fibers than would a perpendicular segment with parallel fibers. Since the increasing torque of the myotome increases the degree of slant, the resistance to lateral expansion is reduced by the act of contraction. A third factor in myotome contraction involves the sequence of innervation of the myotomes. Muscle contractions are initiated by nerve impulses originating in the fish brain and traveling caudally in nerves of the spinal cord. From the spinal cord, spinal nerves branch to the lateral myotomes. There is a spinal nerve serving each myotome so each myotome contracts in order from cranial to caudal. All the fibers in each myotome

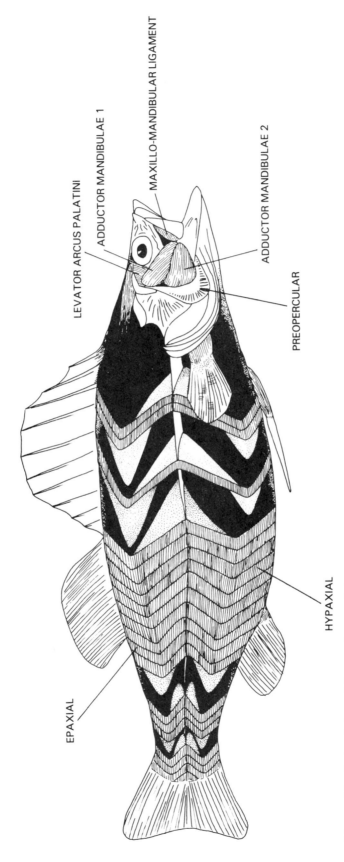

Figure 4.1 Lateral view of the perch musculature.

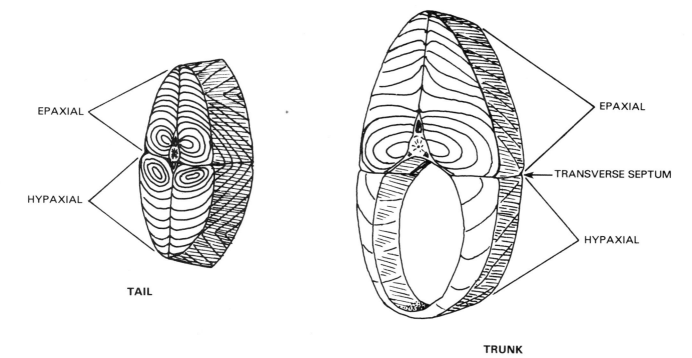

Figure 4.2 Cross section of the trunk and tail regions.

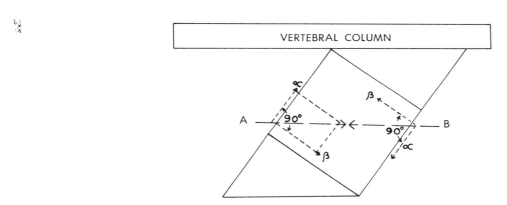

Figure 4.3 Diagram of myomere movement. Since each myosepta is drawn toward the other, A and B indicate the forces in the myomere which draw the myosepta together. These forces are produced by the contraction of the muscle fibers. One of the vectors (β) of each of the forces A or B may be perpendicular to each of the myosepta. The vectors are equal but opposed to similar vectors of other myofibrils and are cancelled out. The other vectors (α) for forces A and B are parallel (90° to the first vectors) to the surface of the myosepta. Since they are not directly opposed, these are the effective forces in the movement of the myomere. Actually the set of vectors tends to torque the myomere in a direction opposite to the set. The β set of vectors are ineffective except at the two extremes of the myomere. The resultant movement at any given level is a torque of the myomere caused by the parallel α vectors.

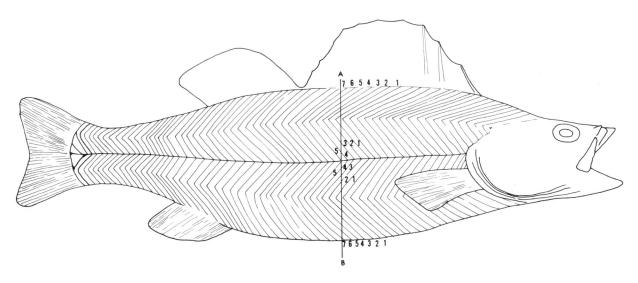

Figure 4.4 Diagram illustrating the gradation of body constriction when slanted myotomes are innervated. (See text for explanation.)

contract simultaneously but the slanted arrangement of the myotomes provides for a smooth increase and decrease in contraction strength at each body level. The arrangement of overlapping myotomes is shown in figure 4.1. If a series of myotomes are numbered as in figure 4.4, and a line is drawn to represent a cross-sectional body level, the number of fibers contracting at a given cross-sectional level is seen to increase gradually. Thus, the first myotome to contract at level A-B is myotome 1A-B followed by 2A-B, 3A-B, 4A-B, 5A-B, 6A-B, and finally 7. As myotome 7 is contracting, myotome 1 is relaxed and 2 is either relaxed or beginning to relax. Thus the peak of contraction is somewhere between the contraction of myotome 2 and myotome 7.

We may conclude, therefore, that the complex arrangement of fish myotomes is due to three selective factors: (1) the ability to torque and thus produce a bending movement; (2) the necessity of the segment to expand laterally as it contracts cranial-caudally and (3) the innervation and overlapping of myotomes to produce a graded increase and decrease in contraction at each cross-sectional level of the body.

It is clear that considerable energy expended during muscle contraction does not result in movement. Expended energy that does not produce work (movement) is released as heat. Muscle contraction is the major source of body heat in all vertebrates. In aquatic *poikilotherms* (poi-KIL-*o*-therms), with little or no body insulation or other heat conservation mechanisms, heat must be produced almost continuously.

Each myotome is divided into dorsal and ventral halves by a transverse septum of connective tissue. Those muscles or myotome segments dorsal to the transverse septum (fig. 4.2) are referred to as *epaxial* (ep-AK-se-al), and those ventral to the septum are *hypaxial* (hi-PAK-se-al). The cranial hypaxial muscles seem to play only a minor role in locomotion. Their main function is probably to support the body viscera.

PREPARATION OF THE SPECIMEN FOR OBSERVATION OF THE HEAD MUSCULATURE

The musculature operating the parts of the upper and lower jaws, the opercular shield, and associated structures is extremely complex. This extreme intricacy is implied by the great number of separate and moveable bones in the fish skull. During movement of the jaws, all of the moveable units, operculum, jaws, and branchial arches must have coordinated actions. The dissection and observation of all these small head muscles is a tedious chore and most of the information gained is applicable only to the species dissected. For this reason, some of the minor or less accessible muscles of the head should be omitted. To expose the head musculature, insert the tip of a scalpel blade (No. 10) under the skin at the dorso-cranial edge of the preopercular bone. Cut the skin carefully along the entire edge of the preopercular. Grasp the cut skin flap with forceps and pull the flaps forward. If you pull strongly enough the cranial attachment will tear loose without damage to the underlying structures. Note that the maxilla and its skin cover form a flap over the caudal part of the lower jaw. This flap may be removed with scissors. A small area of skin just above and caudal to the dorsal tip of the preopercular bone should be cleared away. A pair of blunt forceps will be useful for this chore.

THE HEAD MUSCULATURE

There are two major functions of the head musculature in the perch. The dominant role is feeding and the second important activity is respiration. Each of these functions is extremely complicated and the description here is necessarily a simplification. For a complete description supported by experimental evidence see Osse (1969).

FEEDING

Feeding in the perch is accomplished by "sucking" prey into the mouth, unlike many other fish that bite their prey. The "sucking" movements are performed in two phases. Phase 1 involves protrusion of the jaws and enlargement of the buccal cavity. This expansion takes place in three steps: (1) depress mandible and raise opercula; (2) raise neurocranium and protrude upper jaw, abduct suspensory complex and draw pectoral girdle caudally; (3) draw hyoid ventro-caudally and expand branchiostegal apparatus. Phase 2 compresses the buccal cavity and draws the protruded mouth back in a much slower action. Movements of the head in phase 2 may be divided into 4 events: (1) closing the jaws with the mouth protruded and lowering the opercula; (2) adduction of the suspensorium and opercula, lowering the neurocranium and dorsal movement of the branchiostegal rays; (3) continued adduction of the opercula and suspensoria, forward movement of the hyoid apparatus and pectoral girdle; and (4) completion of adduction of suspensorium and operculum. During phase 2 the pectoral fins are abducted and drawn forward, thus deflecting the current of water flowing from the opercular chambers.

RESPIRATION

Inspiration is principally an abduction of the suspensorium and operculum and an expansion of the branchiostegal apparatus and hyoid. The jaws are opened by movements of the opercular bones that attach to the lower jaw by an *interopercular-mandibular ligament*. The muscles that open the jaws during feeding do not contract during respiration. Expiration begins with closure of the mouth by the two middle portions of the adductor mandibulae. This is followed by adduction of the suspensoria and operculi. The two hyoids and jaws are contracted as the branchiostegal apparatus is compressed to complete expiration.

JAW MUSCLES

NAME	ORIGIN	INSERTION	ACTION
Adductor mandibulae 1 (ah-DUK-tor man-DIB-*u*-lah)	Dorsal half of ventral part of preopercular and hyomandibula.	Dorsal mid-portion of a stout maxillo-mandibular ligament. Ligament extends between articular and dorso-rostral ridge of maxilla. (fig. 4.6)	Active only during feeding. During protrusion of jaws, maintains tension on maxillo-mandibular ligament, thus holding maxilla in place and slowing the movement of maxilla and mandible. Assists closure of jaws after protrusion.
Adductor mandibulae 2	Ventral half of vertical arm of preopercular.	Maxillo-mandibular ligament medial to insertion of adductor mandibulae 1.	Closes mouth and assists adduction of suspensory complex. Active during expiration and feeding.
Adductor mandibulae 3	Dorsal portion of the plate of the preoperculum medial to adductor mandibulae 1 and from portions of hyomandibular, metapterygoid, and symplectic bones.	Maxillo-mandibular ligament medial to insertion of adductor mandibulae 1. Difficult to separate from adductor mandibulae 2. (fig. 4.6)	Closes mouth and assists adduction of suspensory complex. Active during expiration and feeding.
Adductor mandibulae 4 *With a sharp scalpel carefully slice away the lateral portion of the dentary bone to expose the fourth part of the adductor mandibulae.*	Medial, rostral portion of maxillo-mandibular ligament opposite to insertions of the other 3 parts. Tendons from medial rostral surface of preopercular and quadrate.	Meckel's cartilage and dentary bone nearly reaching the symphysis. Beneath intermandibularis.	Closure of jaws during feeding. Not active in respiration.

JAW MUSCLES

NAME	ORIGIN	INSERTION	ACTION
Intermandibularis (in-ter-man-dib-*u*-lar-is)	Muscularly from the rostral medial surface of dentary ventral to Meckel's cartilage near symphysis.	To its fellow in the midline. Lies between the geniohyoideus rostral and caudal. *Spread the rostral geniohyoideus right and left, to observe.*	Draws jaws together; assists in expiration by depressing size of buccal cavity.
Levator arcus palatini (l*e*-V*A*-tor AR-kus pal-ah-T*E*N-i)	Caudal circumorbital bone and dermosphenotic via a large superficial aponeurosis caudal to the orbit.	Lateral surface of hyomandibula and dorsal part of lateral surface of metapterygoid.	Abducts the suspensory complex (palatine, ectopterygoid, metapterygoid, symplectic, preopercular, hyomandibular, quadrate, and entopterygoid) thus enlarging buccal cavity and protruding mouth. Assists depression of lower jaw and moves origin of mandibulae dilator operculi laterally, thus increasing that muscle's advantage.
Adductor arcus palatini (not illustrated) *Remove one eye to see this and the following muscle. Deep to levator arcus palatini.*	Parasphenoid and prootic bones.	Medial surfaces of metapterygoid and hyomandibula.	Adducts the suspensory complex thus depressing buccal cavity.
Adductor hyomandibulae (h*i*-*o*-man-DIB-*u*-lah)	Caudal lateral prootic and ventral surface of pterotic.	Medial surface of opercular process of hyomandibular.	Assists Adductor arcus palatini.
Dilator operculi (d*i*-L*A*-t*or* o-PER-k*u*-l*i*)	Dermosphenotic, hyomandibula pterotic.	Dorsal medial border of opercula.	Mechanical advantage increased by Levator arcus palatini action. Abducts operculum, thus enlarge buccal and opercular cavities and protrudes mouth.
Levator operculi cranialis	Caudal portion of pterotic.	Medial surface of opercula caudal to the hyomandibula-opercula articulation.	Depresses lower jaw and assists in protrusion of mouth.
Levator operculi caudalis	Ventral border of posttemporal.	Medial surface of opercula at caudal part of insertion of Levator operculi cranial.	Assists Levator operculi cranial.
Adductor operculi *Pull the operculum aside to see the medial side and, thus, the adductor operculi.*	Exoccipital and pterotic bones.	Medial surface of opercular ventral to insertion of Levator operculi.	Adducts operculum, thus compressing opercular cavity during expiration.

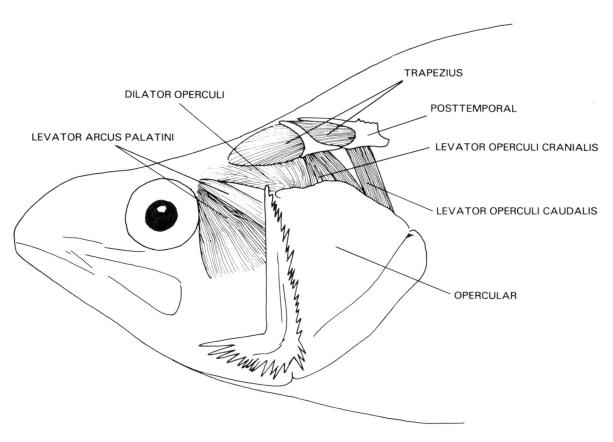

Figure 4.5 Opercular musculature. The adductor mandibulae has been removed (see fig. 4.1).

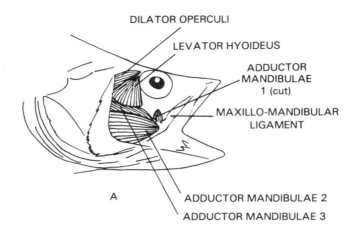

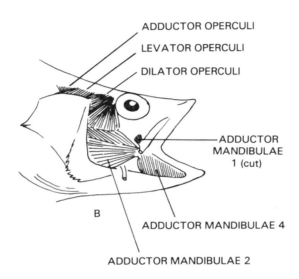

Figure 4.6 Jaw musculature of the perch. A. The adductor mandibulae, part 1 is removed. B. Deep musculature of the jaws.

HYOID MUSCLES

NAME	ORIGIN	INSERTION	ACTION
Geniohyoideus cranialis (je-n*e-o*-h*i*-OYD-*e*-us)	Cranial surface of the ventral half of the first myoseptum.	Ventral edge of cranial region of dentaries by short round tendon. Ventral to Intermandibularis.	Assists in constricting buccal cavity during expiration. Probably draws hyoid apparatus forward.
Geniohyoideus caudalis	Ceratohyals. The muscle is subdivided by the first and second myosepta. A horizontal aponeurotic plate extends forward from the dorsal part of the first myosepta to the dorsal medial cranial surface of the dentary.	Horizontal aponeurotic plate to the medial cranial surface of dentary dorsal to Intermandibularis.	During inspiration, enlarges buccal cavity by abduction of suspensory complex and assists in depression of the floor of the buccal cavity. During expiration assists Geniohyoideus cranial as lower jaw is fixed by adductor mandibulae.

HYOID MUSCLES

NAME	ORIGIN	INSERTION	ACTION
Hyohyoideus inferior (hi-o-hi-OYD-e-us) *Separate the geniohyoideus caudal at the midline to see the hyohyoideus.*	Hypohyal	First and second branchiostegal rays opposite the side of origin.	During inspiration expands branchiostegal rays, thus expanding buccal and opercular cavities.
Hyohyoideus superior (not illustrated)	Dorsal edge of seventh branchiostegal ray dorsal to the branchiostegal membrane.	Medial side of subopercular and opercular.	During expiration constricts branchiostegal rays, thus compressing the opercular cavity.
Hyoideus proprius (not illustrated) (PRO-pre-us)	Dorsal process of ceratohyal.	Lateral surface of hypohyal.	Appears to maintain a tension on the dorsal side of the hyoid apparatus. May assist Geniohyoideus caudal.
Sternohyoideus (ster-no-hi-OYD-e-us)	Ventral cranial surface of cleithrum and the ventral myoseptum continuous with the cleithrum. Two transverse myosepta subdivide the muscle before it inserts.	Lateral surface of urohyal. This muscle is continuous caudally with hypaxial musculature.	Retracts hyoid, draws lower jaw caudally and depresses lower jaw. Does not contract during inspiration.
Trapezius (trah-PEze-us)	Pterotic and epiotic bones dorsal to the dermosphenotic-pterotic ridge.	Ventral edge of lateral surface of posttemporal and dorsal tip of cleithrum.	Draws pectoral girdle dorso cranially, thus drawing origin of Sternohyoideus caudally.

EYE MUSCLES

(Also see page 73, Chart of the Cranial Nerves)

Separate the fascia from the edge of the orbit, pull the eye laterally and cut the muscles away from the skull while leaving as much attached to the eyeball as possible. As the eye is removed note the orientation and locate the optic nerve (fig. 4.11).

NAME	ORIGIN	INSERTION	ACTION
Superior rectus (REK-tus)	Caudal end of parasphenoid (see chap. 3., II., number 1.).	Just caudal of dorsal midpoint of eyeball.	Rotates eye dorsally with superior oblique and caudally with lateral rectus.
Lateral rectus	Caudal end of parasphenoid.	Dorso-caudal border of eyeball.	Rotates eye caudally.
Inferior rectus	Caudal end of parasphenoid ventral to origin of superior rectus.	Caudal to ventral midpoint of eyeball.	Rotates eye ventrally with inferior oblique.
Medial rectus	Middle of horizontal bar of parasphenoid.	Mid-rostral border of eyeball.	Rotates eye forward (rostrally).
Superior oblique (o-BLEK)	Rostral end of parasphenoid.	Just rostral of dorsal midpoint of eyeball.	Rotates eye dorsally with superior rectus.
Inferior oblique	Rostral end of parasphenoid.	Rostral to ventral midpoint of eyeball.	Rotates eye ventrally with inferior rectus.

APPENDICULAR MUSCULATURE

NAME	ORIGIN	INSERTION	ACTION
Pectoral fin adductors	Scapula and coracoid.	Radials and on the base of the lepidotrichia.	Adduct the fin.
Pectoral fin abductors (ab-DUK-tors) (fig. 4.8) *Pull fin laterally to see these fibers.*	Medial surface of the scapula and coracoid.	Medial surface of the radials.	Abduct the fin.
Pectoral fin rotators (not illustrated)	Fin adductors and abductors.	Leading and trailing edges of the fin.	Rotate the fin around the longitudinal axis.
Pelvic fin adductors (fig. 4.7) *Pull fin laterally to see these fibers.*	Medial surface of the pelvic plates.	Medial surface of the fin radials.	Adduct the fin.
Pelvic fin abductors	Medial surface of pelvic plates.	Lateral surface of the fin radials.	Abduct the fin.
Median fins (cranio-dorsal, caudal dorsal and anal fins) (fig. 4.9) *Observe with a dissecting scope.* (often damaged during removal of the skin)	Group 1 (*inclinator dorsalis*): connective tissue at the sides of the fin. Group 2 (cranial-most section called *supracarinalis anterior*): pterygiophores.	Group 1: bases of the lepidotrichia. Group 2: pterygiophores.	Erect the fins.
Caudal fin (fig. 4.10)	Myomeres.	Lepidotrichia.	Operate caudal fin.

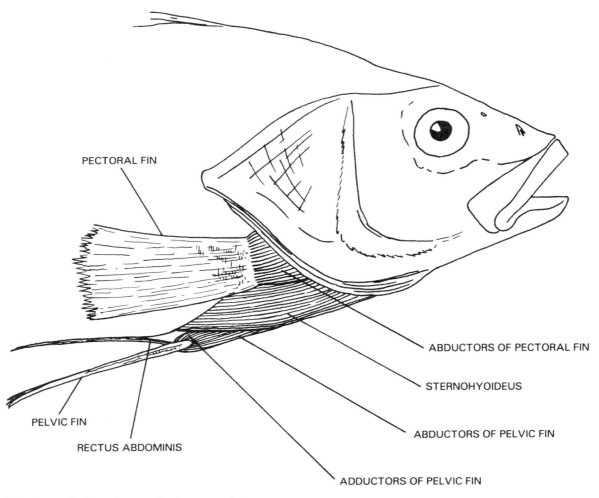

Figure 4.7 Lateral view of appendicular musculature.

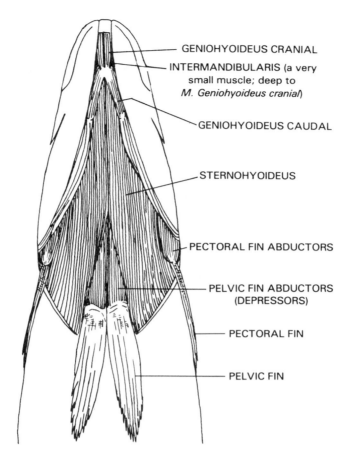

Figure 4.8 Ventral view of jaw musculature.

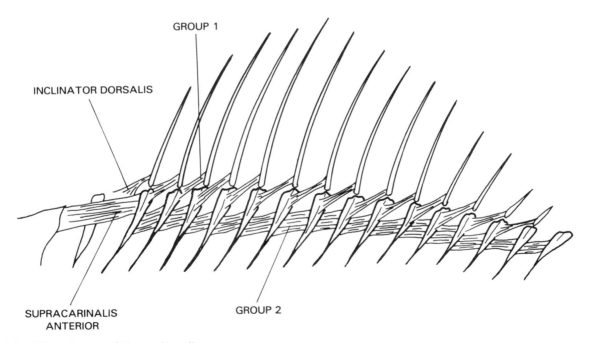

Figure 4.9 Musculature of the median fin.

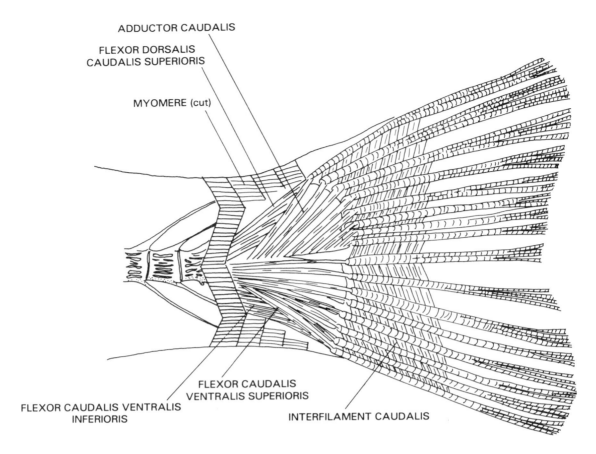

Figure 4.10 Musculature of the tail.

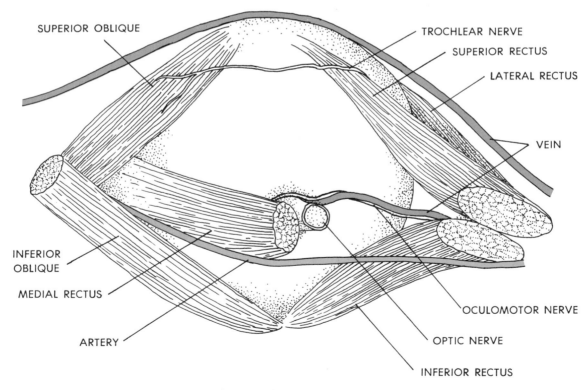

Figure 4.11 Caudal view of the left eyeball of the perch.

SUGGESTED READINGS

Bone, Q. and R. D. Ono. 1982. Systematic implications of innervation patterns in teleost myotomes. *Brevoria* 470:1–23.

Branson, B. A. 1966. Guide to the muscles of bony fishes, excluding some special fibers in Silurooids and a few other. *Turtox News* 44(4):98–102.

Daan, S. and T. Belterman. 1968. Lateral bending in locomotion of some lower tetrapods. *Koninkl. Nederl. Akademie van Wetenschappen Proc.* II. 71(3):259.

Lauder, G. V. 1980. The evolution of the jaw adductor musculature in primitive gnathostome fishes. *Breviora* 460:1–10.

Nursall, J. R. 1956. The lateral musculature and the swimming of fish. *Proc. Zool. Soc. Lond.* 126:127–143.

———. 1962. Swimming and the origin of paired appendages. *Amer. Zoologist* 2:127–141.

Osse, J. W. M. 1969. Functional morphology of the head of the perch (*Perca fluviatilis* L.). An electromyographic study. *Netherlands Jour. of Zoology* 19(3):289–392.

Szarski, H. 1964. The functions of myomere folding in aquatic vertebrates. *Bull. de L'Acad. Polonaise des Sciences.* Cl.II.Vol.XII (7):305–306.

Van Hasslet, M. J. F. M. 1979. Morphology and movements of the jaw apparatus in some Labrinae, Pisces, Perciformes. *Netherlands Jour. of Zoology* 29(1):52–108.

Wainwright, P. C. 1989. Functional morphology of the pharyngeal jaw apparatus in Perciform fishes: An experimental analysis. *Jour. Morph.* 200:231–245.

Willemse, J. J. 1959. The way in which flexures of the body are caused by muscular contractions. *Koninkl. Nederl. Akad. Wetensch. Amsterdam Proc.* 62:589–593.

———. 1966. Functional anatomy of the myosepta in fishes. *Koninkl. Nederl. Akad. Wetensch. Amsterdam Proc.* Series C., 69(1):58–63.

Chapter 5
Mouth, Pharynx, and Respiratory System

PREPARATION OF THE SPECIMEN FOR OBSERVATION OF THE ORAL CAVITY AND PHARYNX

Place one blade of your scissors into the mouth and cut each lateral corner back about one inch. Be careful not to damage the gills or tongue as you make these cuts.

Mouth and Pharynx (fig. 5.1).

1. The *mouth* of the perch is bordered by the premaxilla and maxilla above and the dentary below. Teeth are present only on the premaxilla and dentary. The maxilla serves only as a lateral wall, channelling water and prey into the oral cavity.
2. The *oral cavity* (*O*-ral) is a shallow vestibule arbitrarily separated from the more caudal pharynx.

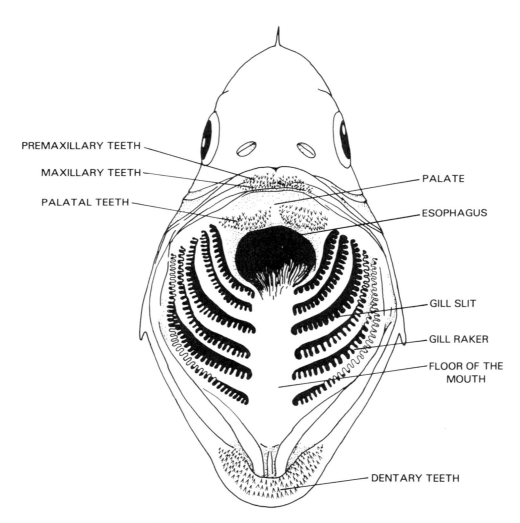

Figure 5.1 Oral cavity and pharynx of the perch.

The gill rakers and gill slits may be considered parts of the pharynx (see number 4 that follows) so the oral cavity is between the first row of gill rakers and the mouth.

3. The *palate* (PAL-at) forms a relatively solid roof for the oral cavity and pharynx. Palatal teeth are present on the central vomer and lateral palatine bones, thus forming a cranially directed arch of cranial palatal teeth.
4. The *pharynx* (FAR-inks) contains the branchial complex of hyoid apparatus and gill arches (see chapter 3., page 20). Pharyngeal teeth are present on both the roof and floor of the pharynx. The pharyngobranchial bones serve as a base for the pharyngeal teeth in the roof of the pharynx, and a *dentigerous plate* (den-TIJ-er-us) attached to the medial portion of the most caudal branchial arch bears the teeth of the pharynx floor. The pharyngeal teeth are in the most caudal part of the pharynx and probably serve to hold the struggling prey just prior to swallowing. The act of swallowing will constrict the pharynx and the caudally directed pharyngeal teeth will then assist in driving the prey into the esophagus.
5. Five pair of *gill slits* perforate the pharyngeal wall. The first slit is between the last branchiostegal ray and the first branchial arch. The last slit is between the last branchial arch and the lateral portion of the dentigerous plate. The three middle gill slits are between the successive branchial arches. The mid-ventral floor of the pharynx is composed of basibranchials based cranially on the hypohyal.

The Respiratory System

1. The *gills* are attached to each of the four pairs of branchial arches. The gills are on the aboral surface of the branchial arch and the *gill rakers* (RA-kers) are on the oral surface. Laterally, the gills are covered by the opercular shield (see chapter 3., page 17). Ventrally, the "fanlike" branchiostegal apparatus completes the opercular chamber.
2. Each gill arch has two sets of *filaments* (FIL-a-ments), one set extending into the slit behind the arch. The cranial set of filaments are *posttrematic* (tre-MA-tic) (at the caudal border of the slit) and the caudal filaments are pretrematic (at the cranial border of the slit). Since all of the gills have a double set of filaments they are known as *holobranchs* (HOL-o-branks). The holobranch filaments extending into one gill slit (fig. 5.2) are known as *hemibranchs* (HEM-e-branks) (one-half of a holobranch) and nonfunctional gill filaments (as in the spiracle of the shark) are called *pseudobranchs* (SU-do-branks).

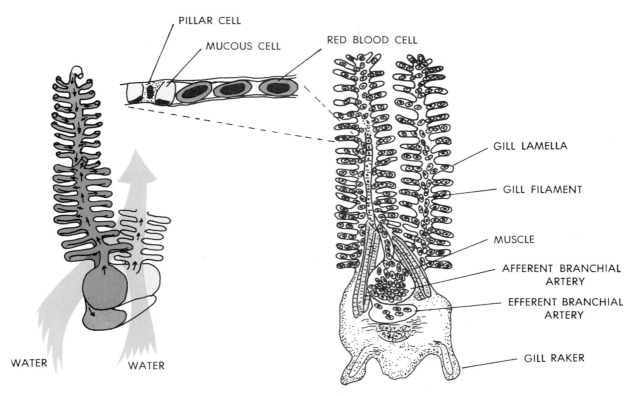

Figure 5.2 Diagram of the gills and gill circulation. Note that blood and water flow in opposite directions. Water flow is indicated by the arrows.

Each gill filament is finely subdivided into folds called *lamellae* (la-MEL-*e*). Gill capillaries are located inside the lamellae and the exchange of gases, oxygen, and carbon dioxide takes place between the blood and water through the wall of the lamellae.

Perch have finely divided gills with as many as 31 lamellae per square millimeter. The lamellae are very small (only 10 microns thick) and have thin walls so the distance between blood and water is less than one micron. There is a square centimeter of gill surface for every 8 grams of body weight in the perch.

THE MECHANISM OF RESPIRATION

Water passes into the fish mouth to the pharynx and from the pharynx through the gill slits into the opercular chamber (fig. 5.1). A backflow of water, out of the mouth, is prevented by the *oral valves*. The water passes out of the opercular chamber at the caudal free edge of the opercular flap. The opercular chamber is closed when the free edge of the operculum is held against the body.

Actually, the flow of water is virtually continuous over the perch's gills. The mechanism that "pumps" this water is analogous to a continuous flow pump.

Contraction of the sternohyoideus and geniohyoideus open the mouth and draw the floor of the pharynx downward. Contraction of the levator hyoideus helps to expand the buccal cavity laterally and the dilator operculi does the same for the opercular chamber. Branchiostegal muscles (the hyohyoideus) separate the branchiostegal rays, thus expanding the ventral portion of the opercular chamber.

The adductor mandibulae closes the mouth and compresses the lips. The relaxation of the sternohyoideus and contraction of the geniohyoideus raises the hyoid. The levator arcus palatini compresses the wall of the oral cavity and the adductor operculi compresses the opercular chamber.

As water passes over the gill filaments, oxygen (suspended) is carried to the filament and carbon dioxide is carried away. Inside the filament, the blood vessels are so arranged that the blood flow is opposite to the water flow over the gills (see figure 5.2). This arrangement is known as *countercurrent exchange*. Thus the oxygen is carried away immediately as it enters the bloodstream and carbon dioxide is carried (by the blood) to the filament.

SUGGESTED READINGS

Boland, E. J. and K. R. Olson. 1979. Vascular organization of the catfish gill filament. *Cell Tissue Res.* 198:487–500.

Hughs, G. M. 1981. The dimensions of fish gills in relation to their function. *Jour. Experimental Biology* 45:177–195.

Olson, K. R. 1981. Morphology and vascular anatomy of the gills of a primitive air breathing fish, the bowfin (*Amia calva*). *Cell Tissue Res.* 218:499–517.

Osse, J. W. M. 1969. Functional morphology of the head of the perch (*Perca fluviatilis*), an electromyographic study. *Netherlands Jour. of Zoology* 19(3):289–391.

Satchell, G. H. 1968. A neurological basis for the coordination of swimming with respiration in fish. *Comp. Biochem. Physiol.* 27:835–841.

Steen, J. B. and A. Druysse. 1964. The respiratory function of teleostean gills. *Comp. Biochem. Physiol.* 12:127–142.

Steen, J. B. and T. Berg. 1966. The gills of two species of haemoglobin free fishes compared to those of other teleosts, with a note on severe anaemia in an eel. *Comp. Biochem. and Physiol.* 18:517–526.

Wiley, E. O. 1979. Ventral gill arch muscles and the interrelationships of gnathostomes with a new classification of the vertebrata. *Zool. J. Linn. Soc.* 67(2):149–180.

Chapter 6
Body Cavities and Viscera

PREPARATION OF THE SPECIMEN FOR OBSERVATION OF THE VISCERA

In order to study the internal organs it will be necessary to cut out a portion of the trunk wall. Make an incision along the ventral trunk wall beginning just in front of the anal opening and continue the incision cranially to the pelvic girdle. At the cranial and again at the caudal end of the ventral incision, make cuts dorsally, ending your incisions just ventral to the lateral line. Be very cautious in making your incisions. A deep cut may destroy some of the underlying organs. If you are sufficiently careful the parietal peritoneum will remain with the coelom rather than with the body wall that you remove. Layers of fat will be found both retroperitoneal and in the mesenteries. Carefully remove the fat with forceps. In spite of your care you will tear and destroy some of the serous membranes but this cannot be avoided. The pericardial cavity may be exposed by cutting transversely across the sternohyoideus muscles approximately one inch cranial to the pelvic fins. Continue this incision dorsally to a point cranial and dorsal to the pectoral fin. Again, do not cut through the lateral line. Next, make a mid-ventral incision caudally to the trunk incision you made earlier. The pericardial membranes should adhere to the flap you have cut. With forceps, work the pericardia loose as you raise the flap.

MESENTERIES AND COELOMIC CAVITIES

The body of the perch surrounds a large cavity (the *peritoneal cavity*) (per-i-to-N*E*-al) containing the *visceral* (VIS-er-al) organs (digestive tract and glands, reproductive organs, and swim bladder). Another cavity, cranial to the peritoneal cavity, containing the heart, is called the *pericardial* (per-i-KAR-de-al) cavity.

The combined pericardial and peritoneal cavities form embryonically as double cavities extending the length (cranial-caudal) of the perch's trunk. Each cavity is lined with a membrane called a *serous* (S*E*-rus) membrane. The cavities are called *coelomic cavities* (se-LOM-ik) and the fluid that fills them is *coelomic fluid*. Much of the viscera is suspended between the fluid-filled sacs, and the serous membranes of the coelomic cavities meet in the midline above and below the viscera they support. These membranes are called, respectively, *dorsal* and *ventral mesenteries* (MES-en-ter-es).

As embryonic development continues, the pericardial and peritoneal cavities become separated from one another and the mesenteries of the pericardial cavity disappear. In the peritoneal cavity most of the ventral mesentery also disappears. At the cranial end of the peritoneal cavity a remnant of the ventral mesentery, the *falciform ligament* (FAL-si-form), attaches the liver to the ventral body wall and continues around the liver and attaches the liver to the stomach where it is called the *gastrohepatic ligament* (gas-tro-he-PAT-ik).

In the adult perch the serous membrane encircling the viscera is the *visceral peritoneum*. The single-layered membrane lining the peritoneal cavity is the *parietal peritoneum* (pah-R*I*-e-tal). In the pericardial cavity the comparable membranes are the *parietal* and *visceral pericardia* (per-i-KAR-de-a).

This arrangement permits the visceral organs to move about and provides lubrication at the same time. This is especially important for moving organs such as the heart, stomach, and intestines. In addition, the coelomic fluid serves as a medium for the passages of wastes and, in the female, for the passage of eggs. Structures that are not suspended by mesenteries in the trunk cavity are designated as *retroperitoneal* (re-tro-per-i-to-N*E*-al). This means that the structures are behind (*retro*) the peritoneum.

The Trunk Viscera (fig. 6.1)

The body cavity is filled with the various visceral structures: *digestive tract, gonads (testes or ovaries), spleen, kidneys,* and *swim bladder*. All of these visceral structures are covered by thick layers of fat contained by the serous membranes. The digestive tract includes the gut tube and the accessory digestive glands such as *liver* and *pancreas* (see chapter 7).

1. The *spleen* is an elongated structure lying on the caudo-dorsal surface of the stomach. This structure is concerned with the production and maintenance of blood cells (see chapter 8).
2. The *gonads* (G*O*-nads) occupy the space in the trunk cavity dorsal to the intestine and caudal to the stomach (see chapter 9). In addition to those structures contained in the peritoneal cavity, the kidneys and the swim bladder are located retroperitoneally, that is, behind the peritoneum.

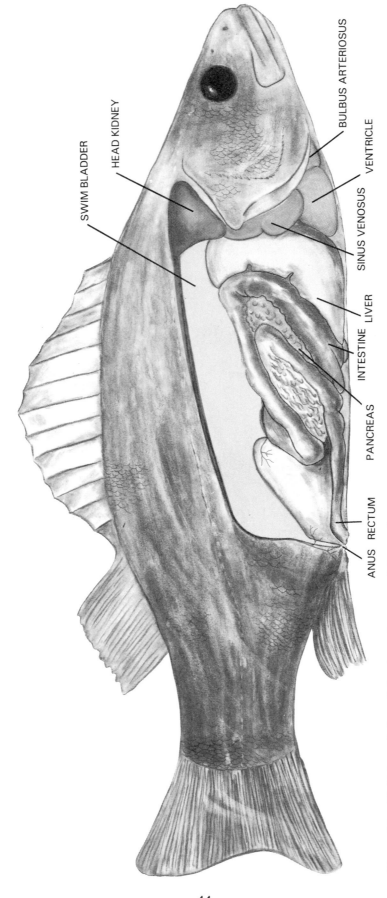

Figure 6.1 Right lateral view of the visceral organs of the perch.

3. The *swim bladder* of some bony fishes is a saclike organ containing a gas. There is no swim bladder in the cartilaginous fishes and only a little more than half of the teleosts have this structure, although it is usually present in the larvae of bony fishes. The swim bladder usually serves as a hydrostatic organ allowing the fish to maintain a specific depth without exerting energy. Fishes without a swim bladder must swim continuously in order to maintain depth. The perch has a swim bladder but the duct connecting the bladder to the pharynx is closed so air cannot be obtained or removed through the mouth. Those fish with an open air (pneumatic) duct to the bladder are called *physostomes* (FI-so-stomes). Those with a closed swim bladder are called *physoclists* (FI-so-klists). Physoclists, such as the perch, with separate gas-secreting and gas-absorbing areas in the swim bladder are termed *euphysoclists* (U-fi-so-klists).

   The gas in the perch swim bladder varies in its composition from 12 percent oxygen and 1.4 percent carbon dioxide to 25 percent oxygen and 2.9 percent carbon dioxide.

   Since buoyancy is greater in salt water than in fresh water, freshwater fish like the perch have a larger swim bladder than do the marine fishes.

   The swim bladder has an outer layer of fibrous connective tissue that serves as a barrier to gas loss and a middle layer of jellylike material. The innermost layer of the swim bladder is the mucosa and it is of two types in the perch. The caudal absorbing chamber is lined with thin flat cells. The rostral gas gland epithelium includes a cuboidal cell epithelium, a middle layer of blood capillaries and smooth muscles, and an outer layer of collagen fibers.

4. The *kidneys* (KID-nes) adhere closely to the subvertebral musculature along the dorsal midline (see chapter 9).

SUGGESTED READINGS

Fänge, Ragnar. 1966. The physiology of the swim bladder. *Physiological Reviews* 46(2):299–322.

Steen, J. B. 1970. The swim bladder as a hydrostatic organ. In *Fish Physiology*. eds. W. S. Hoar and D. J. Randall, vol. IV, pp.413–443. New York: Academic Press.

Chapter 7
The Digestive System

If you have not yet opened the body cavity of your specimen see the beginning of chapter 6 on the preparation of the specimen for the observation of the viscera.

The digestive tract includes the gut tube and the accessory digestive glands, *liver* (LIV-er) and *pancreas* (PAN-kre-as) (see fig. 7.1). The perch is a *carnivore* (KAR-ni-vor) or flesh eater and consequently has a very short intestine. *Herbivorous* (er-BI-vor-us) (plant-eating) fish have a long intestine which is several times the length of the body. The total length of the perch intestine is less than the length of the body. This correlation between gut length and feeding habits is based on the supposition that a greater intestinal surface is necessary for digestion of plant material.

HISTOLOGY OF THE DIGESTIVE TUBE

The digestive tube of vertebrates has several layers of tissues surrounding the central cavity (lumen). Fish have most of these layers although there are differences in the specific structure of the layers between fish and other vertebrates. The following layers (see fig. 7.2) may be observed in a cross section of a fish intestine; (1) *mucosal epithelium* (myoo-KO-sal), a single layer of cells surrounding the *lumen* (LOO-men). The cells adjacent to the lumen are columnar and frequently ciliated but in the stomach pits extend inward from the surface and these pits are lined with cuboidal cells. Deep in the mucosa (epithelium) is the (2) *submucosa* (SUB-myoo-ko-sa), consisting of connective tissue and blood vessels. In most vertebrates the submucosa is divided into two portions by a thin layer of circular muscles called the (3) *muscularis mucosa*. In fishes the muscularis mucosa is variable. It may occur as a few scattered cells in some fish or as a distinct layer in others. Even when the muscularis mucosa occurs as a distinct layer it may not separate the submucosa into two parts as it does in terrestrial vertebrates. When the submucosa is divided, the portion between the mucosa and the muscularis mucosa is the *tunica propria* (PRO-pre-a) and only the portion external to the muscularis mucosa is referred to as the submucosa. In parts of the perch digestive tract the muscularis mucosa is often adjacent to the mucosal epithelium so only blood channels with blood lie between the muscularis and the epithelium. These blood channels are really insufficient to be termed a tunica propria. External to the submucosa is a (4) *tunica muscularis* consisting of an innermost layer of circular muscles and an outer most layer of longitudinal fibers. The muscle fibers of the tunica muscularis may be striated, smooth, or a mixture of smooth and striated fibers.

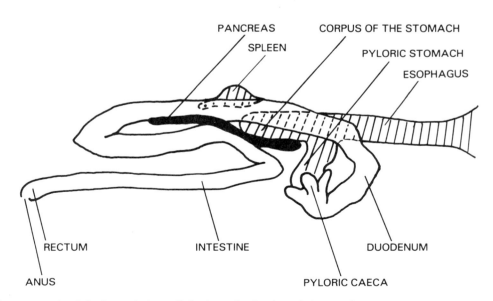

Figure 7.1 Diagrammatic right lateral view of the intestinal tract of the perch.

The digestive tube is surrounded by a thin membrane, the (5) *serosa* (se-RO-sa), throughout most of its length. The serosa is continuous with the double-layered mesentery on its dorsal border.

1. The *esophagus* (e-SOF-a-gus), at the extreme cranial end of the body cavity, is a short, straight tube leading from the mouth to the stomach. Externally the separation between the esophagus and stomach is not evident but internally the stomach wall is thicker and has several longitudinal folds that are not present in the esophagus. There is no *sphincter* (SFINK-ter) or valve at the esophageal-gastric junction but the outer circular muscles form a complete layer at the outer border of the esophagus while the longitudinal muscle fibers are arranged in bundles in the submucosa. The outer *circular muscularis* is composed of both striated and smooth muscle fibers while the longitudinal fibers are all striated. The lining of the esophagus has cuboidal (square) or columnar *mucous cells* producing mucus (slime) to lubricate the food.
2. The *stomach* (STUM-ak) of the perch is separable into a *corpus* (KOR-pus) and a *pyloric* (pi-LOR-ik) stomach. The corpus is a straight portion and receives the esophagus cranially while ending caudally as a blind sack. The pyloric portion extends laterally to meet the proximal end of the intestine. There is a *pyloric sphincter* between the pyloric stomach and the duodenum (some authors have used the term "pylorus" for the pyloric portion of the stomach but pylorus is the name of the passageway between the stomach and the intestine). The separation between corpus and pyloric stomach is based on histological differences between the two regions. The corpus contains unbranched, tubular gastric pits lined by a single gastric gland cell type producing both enzymes and hydrochloric acid. The corpus has a reduction in the number and size of gastric glands and there are no glands in the pyloric stomach. The circular muscle in the wall of the pyloric stomach is twice as thick as the circular muscle in the wall of the corpus (figs. 7.2 and 7.3). *This difference in thickness may also be seen in a gross dissection if the stomach is cut open longitudinally with a sharp scalpel or single-edged razor blade.*

HISTOLOGY OF THE STOMACH WALL

The stomach wall is usually thin with little or no submucosa. The tunica propria is also sparse but appears in longitudinal folds of the epithelium. There are fewer folds and more submucosa at the pyloric end of the stomach. *Gastric glands* occur in pits in the corpus. The pits are lined with cuboidal cells while the lining of the corpus proper consists of columnar cells. The columnar cells lining the stomach are often mucous-secreting cells that discharge a continuous flow of mucus to protect the lining of

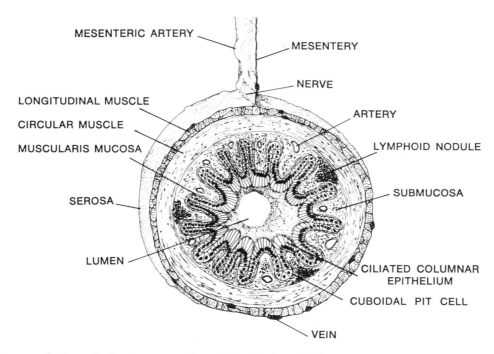

Figure 7.2 Diagram of a hypothetical cross section of the fish intestinal tract illustrating features of several regions of the tract.

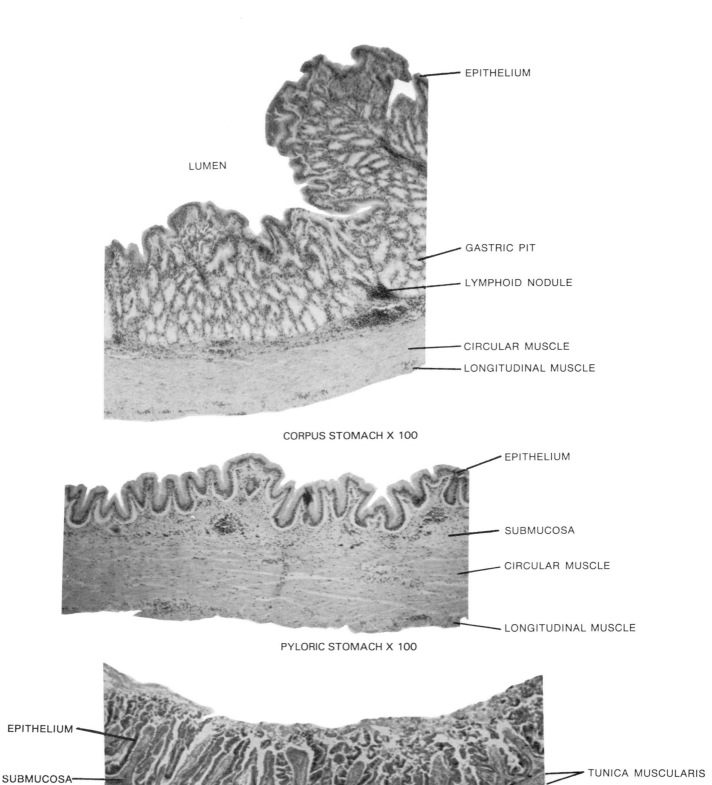

Figure 7.3 Photomicrographs of cross sections of portions of the corpus and pyloric regions of the stomach and of the small intestine × 100.

the stomach from the hydrochloric acid and pepsin produced by the gastric glands of the pits. Cells in the neck of the gastric pits may also be mucous-producing but the deeper cells are acid- and enzyme-producing cells. Mucous-producing cells occur in the pyloric portion but there are no gastric pits in this region.

1. Three blind pouches, the *pyloric caeca* (SE-ka), branch from the gut tube of the pyloric-intestinal juncture. The epithelium lining the caeca is the same as that lining the proximal intestine, but caeca appear to function as part of the stomach by allowing for the ingestion of prey (especially fish) that are longer than the combined corpus and pyloric stomachs. Perch have a cecal stomach by reason of having "intestinal" ceca. Some fish have a spindle-shaped extension of the fore gut called a *rectilinear stomach* (rek-ti-LIN-e-ar); others have a *siphonal stomach* (si-FON-al) or none at all. The absence of a stomach in fish is considered a secondary loss.
2. The *duodenum* (doo-o-DE-num) forms an S-shaped loop between the pyloric stomach and the *intestine* (in-TES-tin). The pancreatic and hepatic ducts enter the duodenum near the intestinal caeca. The intestine proper (fig. 7.4) is separated into proximal and distal regions based on histological differences. The intestine is uniform in diameter throughout its entire length. Consequently, there is no external separation between the proximal and distal intestines but when the intestine is cut longitudinally and viewed internally, the *rectum* (REK-tum) of the perch may be identified by its thicker wall. In addition, the perch has a circular, flaplike *intestino-rectal valve* directed toward the proximal intestine. The muscular rectum terminates in an *anus* (A-nus). The adult perch does not have a cloaca.

HISTOLOGY OF THE INTESTINE

The intestine of carnivorous fishes like the perch is a relatively short, simple tube with shallow longitudinal folds and few *villi* (VIL-i). The lining of the intestine is a simple columnar epithelium with interspersed goblet cells and mucous-secreting cells. There are no submucosal glands or crypts in the fish intestine (figs. 7.3 and 7.4). The muscularis externa is much thinner than that of the stomach (fig. 7.3).

1. The *liver* (LIV-er) is situated just cranial and dorsal to the stomach. Portions of the liver cover the cranial diverticulum and the ventral stomach. The liver is drained by a system of tubules (canaliculi and ductules) into an expanded *gall bladder* (gawl). The gall bladder, in turn, opens by several (4 to 6) *hepatic ducts* (he-PAT-ik) through the dorso-cranial wall of the duodenum caudal to the pyloric caeca.
2. The *pancreas* (PAN-kre-as) is found in the first loop of the proximal intestine. In many fishes the pancreas is associated with the hepatic portal vein and its branches. Consequently much of the pancreas may be scattered throughout the body cavity or embedded in the liver. The *pancreatic ducts* drain into the pancreas and empty to the proximal intestine.

HISTOLOGY OF THE LIVER AND PANCREAS

The liver and pancreas are closely associated in fish (fig. 7.5). The pancreas is located on the hepatic portal vein and usually this is at the point of entrance of the hepatic portal to the liver. This arrangement alters the typical portal triad in which the *portal vein* (POR-tal), *bile duct* (bil), and *hepatic artery* are close together in the mammalian liver. In the fish the parts of the triad are

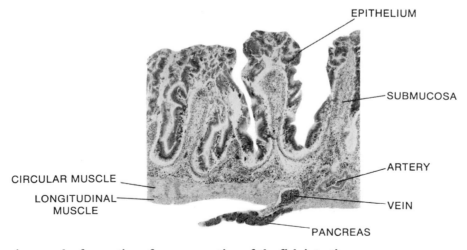

Figure 7.4 Photomicrograph of a portion of a cross section of the fish intestine × 300.

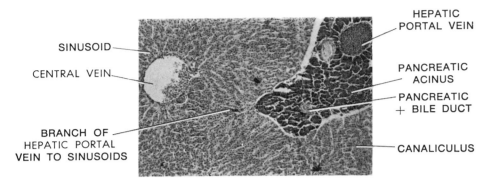

Figure 7.5 Photomicrograph of a portion of the fish liver and pancreas × 500.

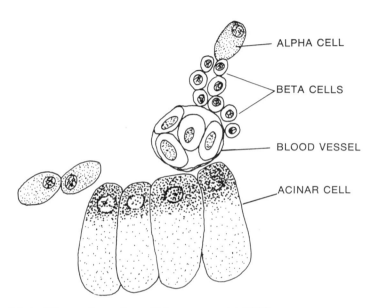

Figure 7.6 Drawing of an islet of Langerhans from fish pancreas × 1000.

widely separated by intervening pancreatic *acini* (AS-i-ni). The fish liver lobules are not well defined but their approximate limits may be estimated as being halfway between two *central veins*. The bile duct is composed of cuboidal or columnar cells surrounding the lumen and situated on a basement membrane. Surrounding this tube is a thick, dense bundle of collagen so that the bile duct appears as a tube within a tube in cross section. The bile duct of fish is usually located in the space between pancreatic and liver parenchyma but is sometimes found in the parenchyma proper. The fish pancreas is illustrated in figure 7.5 with the liver. The cells of the pancreas are arranged in balls or acini adjacent to blood vessels and may occur with the liver or in the mesentery adjacent to the duodenal region of the intestine. The cells are cuboidal or columnar. The bile ducts of the liver are ultimately shared with the pancreas but the pancreatic acini are drained directly into transparent *pancreatic ducts* that resemble small blood vessels. The pancreatic ducts are often filled with pancreatic juice that stains a light pink from the stain used for these sections. The absence of red blood cells will help to distinguish them from small blood vessels.

Clumps of endocrine cells called *islets of Langerhans* (LAHNG-er-hanz), will be seen in most sections. These islets may be associated with small blood vessels in the pancreas and lack the definite boundaries of acini, or they may occur as discrete "islands" of pale cells dispersed between the dark-staining acinar bundles of the pancreas. The islets of Langerhans in the fish have at least two types of cells that may be distinguished in these sections—alpha cells are large with a light pink, granular cytoplasm and the beta cells are smaller and have a clear cytoplasm (see fig. 7.6). The beta cells are thought to produce *insulin* (IN-su-lin).

SUGGESTED READINGS

Hirji, K. N. 1988. Observations on the histology and histochemistry of the esophagus of the perch, *Perca fluviatilis*. *Jour. Fish. Biol.* 22(2):145–152.

Hoshika, K. 1981. The fine structure of esophago-gastric junction 3. Study of esophago-gastric junction in various vertebrates excluding the mammals. *J. Clin. Electron Microsc.* 14(1–2):135–150.

Langer, M. 1979. Histological investigations on the liver of teleosts 3. The system of biliary pathways. *Z. Mikroskopic Anat. Forsch.* 93(6):1105–1136.

Meister, M. F., W. Humbert, R. Kirsch, and B. Vivien-Roels. 1983. Structure and ultrastructure of the esophagus in sea water and freshwater teleosts, Pisces. *Zoomorphology* 102(1):33–52.

Persson, L. 1981. The effects of temperature and meal size on the rate of gastric evacuation in perch, *Perca fluviatilis*, fed on fish larvae. *Freshwater Biol.* 11(2):131–138.

Reifel, C. W. and A. A. Travill. 1977. Structure and carbohydrate histochemistry of the esophagus in ten teleostean species. *Jour. Morpho.* 152:303–314.

———. 1978. Structure and carbohydrate histochemistry of the stomach in eight species of teleosts. *Jour. Morpho.* 158(2):155–168.

———. 1979. Structure and carbohydrate histochemistry of the intestine in ten teleostean species. *Jour. Morpho.* 162(3):343–360.

Sakano, E. and H. Fujita. 1982. Comparative aspects on the fine structure of the teleost liver. *Okajimas Folia Anat. Jpn.* 58(4–5):501–520.

Chapter 8
The Circulatory System

PREPARATION OF THE SPECIMEN FOR OBSERVATION OF THE CIRCULATORY SYSTEM

A study of the blood vessels will require an especially careful dissection. Many structures will need to be destroyed in order to expose the vessels. Be certain therefore that you will not need to review the musculature near the head or body cavity before beginning this dissection. The muscles will be relatively brittle and may be removed with forceps and scalpel.

If the circulatory system of the perch is singly injected, only the branches associated with the dorsal aorta will receive the injection. The venous system, ventral aorta, afferent arteries, and vessels to the head will not be injected. Some of these vessels may be studied without the benefits of an injection but most of the venous system will probably be impossible to find.

Double-injected specimens will have the dorsal aorta and branches from the dorsal aorta filled with red latex; the veins will contain blue latex. The ventral aorta will not be injected. Triple-injected specimens will have the same arrangement as double-injected specimens except the ventral aorta and ventricle of the heart will be filled with yellow latex.

The renal portal veins that drain most of the caudal venous blood and the hepatic portal vessels draining the visceral organs into the liver will not be injected. For this reason the muscles and viscera will appear to have an arterial supply but no venous drainage.

THE HEART AND BRANCHIAL CIRCULATION

The Heart (fig. 8.1)

As was mentioned earlier, the *pericardial cavity* is originally formed as a cranial continuation of the caudal peritoneal cavity. The membrane on the surface of the heart is the *visceral pericardium* and that on the wall of the cavity is the *parietal pericardium*. The pericardial mesenteries disappear at a very early embryonic stage. The heavy membrane separating the pericardial and peritoneal cavities is the combined caudal parietal pericardium and cranial parietal peritoneum. These combined membranes are termed the *transverse septum* (SEP-tum). The heart of the perch has four chambers: *sinus venosus*, *atrium*, *ventricle*, and *bulbus arteriosus*. These chambers are arranged, from caudal to rostral (as listed) so blood passes from one chamber to the other in that order.

1. The *sinus venosus* (SI-nus ve-NO-sus) receives blood from the right and left *common cardinal veins* (KAR-di-nal) or *ducts of Cuvier* (KOO-ve-a) and paired *hepatic veins* from the liver. In the perch the sinus venosus is an inflated vessel, enlarged slightly from the diameter of the common cardinal veins, and opens directly into the large atrium.

2. The *atrium* (A-tre-um) is a single, large chamber extending on both sides of the ventricle. On the right side of the ventricle a portion of the atrium extends as far forward as the ventral aorta. There is no discernible valve between the sinus venosus and atrium but there is a very efficient bicuspid valve between the atrium and the ventricle. This valve is actually a part of the ventricle rather than the atrium and it may best be seen on the dorsal side of the ventricle if the atrium is carefully removed with a pair of small forceps. The valve is very similar in operation to the mammalian semilunar valve. As the ventricle contracts, the rostral and caudal *cusps* (kusps) of the valve fill up on the ventricular side and occlude the opening. As the ventricle relaxes, blood flows from the atrium into the ventricle.

3. The *ventricle* (VEN-tri-kl) is a thick muscular structure with a single internal chamber. Rostrally, the ventricle has a small opening into the bulbus arteriosus. The opening is protected from back flow by an elongated valve that extends into the bulbus and is attached by muscular slips to the inner wall of the bulbus. If the bulbus is carefully torn loose from the ventricle the folds of the valve will be seen adhering to the ventricle. Some of the bulbar muscular slips will be torn loose from the bulbus and will be seen attached to the bulbar surface of the valves.

4. The *bulbus arteriosus* (BUL-bus ar-te-re-O-sus) is the enlarged base of the ventral aorta. More primitive vertebrates may have a *conus arteriosus* (KO-nus) in this position. A conus is capable of muscular pulsation but the bulbus arteriosus is not.

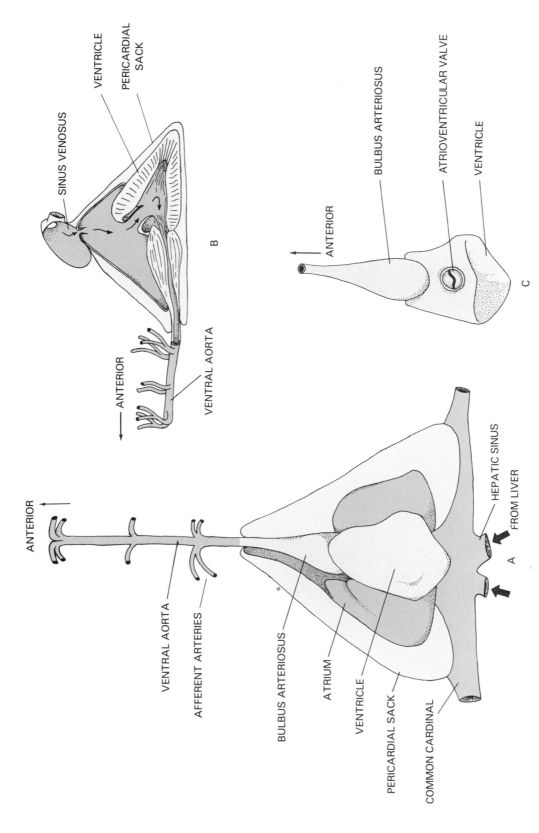

Figure 8.1 The heart and associated vessels of the perch. A. Ventral view. B. Sagittal section of the heart. C. Dorsal view of ventricle and bulbus arteriosus with atrium removed.

According to some authors, the muscle slips and the valvular cusps they adhere to (when the bulbus is separated from the ventricle) together represent the conus arteriosus. This view is not accepted by all anatomists.

In general, the circulation of blood in the teleost is from the heart to the ventral aorta, through the gills and gill capillaries to the dorsal aorta where the blood is distributed to the systematic circulation. This flow is initiated and maintained by the pressure called systole (SIS-to-le) produced by contraction of the ventricle. The pressure is greatest in the ventral aorta and the flow of blood meets its greatest resistance in the gill capillaries. As a consequence of this resistance to flow, the pressure in the dorsal aorta is less than half the pressure in the ventral aorta.

During contraction of the ventricle, the atrium dilates and blood flows into the atrium from the sinus venosus. The walls of the pericardial cavity are allowed only a very slight movement by the ligaments attaching the pericardium to the adjacent cleithra, coracoids, vertebrae, and the units of the fifth gill arch. This rigidity of the pericardium during ventricular contraction creates a negative pressure in the pericardial cavity that is transmitted to the atrial cavity, sinus venosus, and possibly to the entering veins. This negative pressure is the force responsible for drawing blood into the atrium and aids in keeping the *atrioventricular valve* (a-tre-o-ven-TRIK-u-lar) closed during ventricular systole.

During ventricular *diastole* (di-AS-to-le) (expansion) the atrioventricular valve opens and blood flows into the ventricle from the atrium which contracts at this time. During the initial stage of ventricular contraction, blood continues to flow into the ventricle possibly aided by the expansion of the bulbus arteriosus into the floor of the atrium. As ventricular contraction continues, the negative pressure in the atrium closes the atrioventricular valve and opens the *sinoatrial valve* (si-no-A-tre-al).

The rate of heart contraction is accelerated by sympathetic, adrenergic neuronal pathways that accompany the coronary arteries. The cardiac rate is inhibited by parasympathetic, cholinergic neurons.

The Branchial Circulation (fig. 8.2)

1. The *ventral aorta* (a-OR-ta) receives blood from the bulbus arteriosus and distributes it to the afferent branchial (gill) arches. There are only three

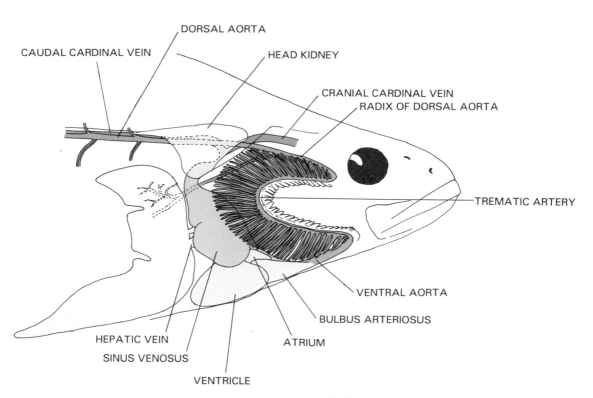

Figure 8.2 Branchial circulation in the perch. Blood enters the gills from the afferent branches of the ventral aorta. Also see the figure 5.2 diagram of gill circulation. Efferent branchial arteries are colored red and enter the trematic artery. Afferent branchial vessels are colored blue and branch from the ventral aorta.

pairs of branches from the ventral aorta. The first and second afferent arches branch directly from the aorta, but the third and fourth afferent arches leave the aorta as a single vessel and branch from this common trunk.

2. The *afferent branchial arteries* extend the entire distance of the gill arch and send capillary branches into the gill filaments where they join branches to the efferent branchial arteries.
3. The *efferent branchial arteries* parallel the afferent arches, each arch receiving a branch from the filaments for every branch sent to the filaments from the afferent arches. The efferent arteries join to form the two roots or radices (a right and a left) of the dorsal aorta. These and the following arteries should be injected with red latex.
4. The *aortic radices* (a-OR-tik RA-di-ces) are the paired roots of the dorsal aorta. The two roots join caudal to the entrance of the last efferent artery to form the dorsal aorta.
5. The *internal carotid arteries* (ka-ROT-id) continue cranially from the first efferent branchial arteries and serve the brain.
6. The *dorsal aorta* extends caudally from its origin (see number 3 above) to the end of the trunk. At the caudal end of the trunk cavity, the vessel continues into the tail through the hemal arches as the *caudal artery*. Branches from the dorsal aorta and caudal artery serve the visceral organs and body muscles of the perch (see the following section).

THE SYSTEMIC ARTERIES

A. The Visceral Arterial Circulation (figs. 8.3 and 8.4)
   1. The *coeliaco-mesenteric artery* (cel-e-AK-o mes-en-TER-ik), the first major cranial branch of the dorsal aorta, continues on the right side of the esophagus and stomach and sends a left gastric artery to the left side of the stomach.
   2. The *left gastric artery* (GAS-trik) branches from the coeliaco-mesenteric and after giving rise to a dorsal gonadal artery the left gastric continues to the left dorsal surface of the stomach.
   3. The *dorsal gonadal artery* (go-NAD-al) extends nearly the full length of the visceral cavity to reach the dorso-rostral center of the gonad.
   4. On the right side of the stomach a *duodenal artery* and small *hepatic arteries* (not illustrated) are the first branches of the coeliaco-mesenteric artery. The duodenal branch also serves the pyloric caeca as well as the proximal portion of the duodenum.
   5. The *right gastric artery* is the third branch of the coeliaco-mesenteric and serves the right dorsal surface of the stomach and sends a pneumatic artery to the air sac.
   6. The *pneumatic artery* (nyoo-MAT-ik) serves the *gas gland* which consists of a *rete mirabile* (RE-te mir-A-bl) of capillaries, arranged so that the direction of blood flow in one capillary is opposite the direction of flow in an adjacent capillary. This *countercurrent* mechanism is apparently important for gas secretion but it is poorly developed or absent in some forms, such as the salmon and herring. The pneumatic artery contains sphincters that may selectively stop the flow of blood to the gas gland.
   7. A large *mesenteric artery* branches from the coeliaco-mesenteric and serves the loop of the intestine. The mesenteric also supplies a *ventrogastric artery* to the ventral surface of the stomach.
   8. The coeliaco-mesenteric artery branches to several small *splenic arteries* (SPLEN-ik) (not illustrated) to the spleen immediately after the mesenteric has branched and then continues to the straight terminal portion of the small intestine as the caudal mesenteric artery.
   9. The *caudal mesenteric artery* sends two or three *ventral gonadal arteries* to the ventral surface of the gonads.

B. The Somatic Arterial Circulation (fig. 8.5)
   At each vertebra the dorsal aorta sends out two pairs of branches. One pair goes dorsally to the interspinous and interpterygiophore muscles and the other pair of arteries serve the muscles between the ribs or hemal arches.
   1. *Interspinal arteries* (in-ter-SPI-nal) are paired branches of the dorsal aorta that fuse and extend dorsally to the dorsal tip of each neural spine. Each pair emerges from the aorta (or caudal artery in the tail) and joins a *supraspinous artery* (soo-pra-SPI-nus) which is an anastomosis running from cranial to caudal at the dorsal tips of the neural arches.
   2. *Interpterygiophore arteries* branch dorsally from the supraspinous artery and serve the interpterygiophore muscles.
   3. *Intercostal arteries* (in-ter-KOS-tal) extend ventrolaterally from the aorta and serve the intercostal muscles. Superficial branches extend laterally from the intercostal arteries and pass through the intermuscular septum, which separates the epaxial and hypaxial muscles.

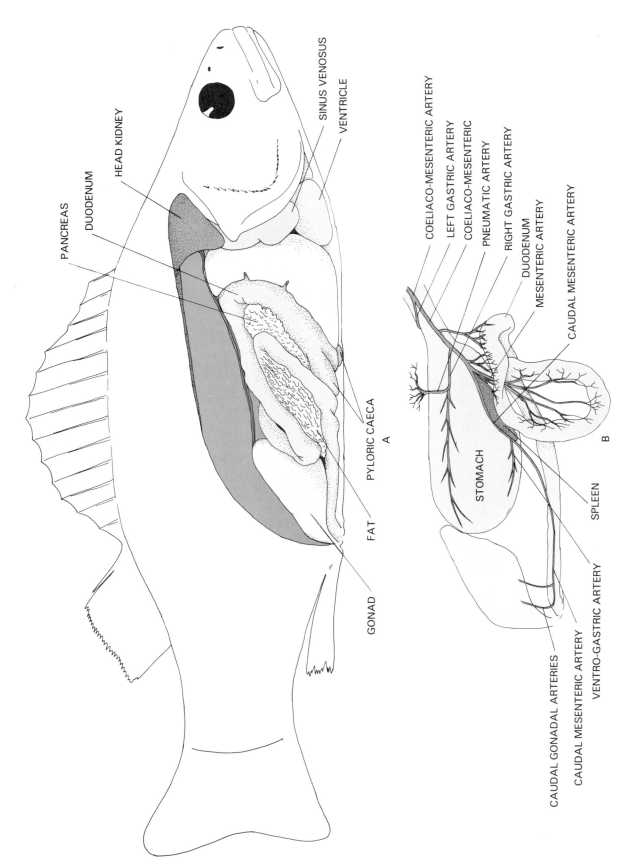

Figure 8.3 Visceral organs and blood vessels on the right side of the perch. A. Illustration of the viscera in natural position with the lateral body wall removed. The blue area indicates the position of the air sack. B. Illustration of the digestive tract with the liver removed and the duodenal loop deflected downward to expose the branches of the coeliaco-mesenteric artery.

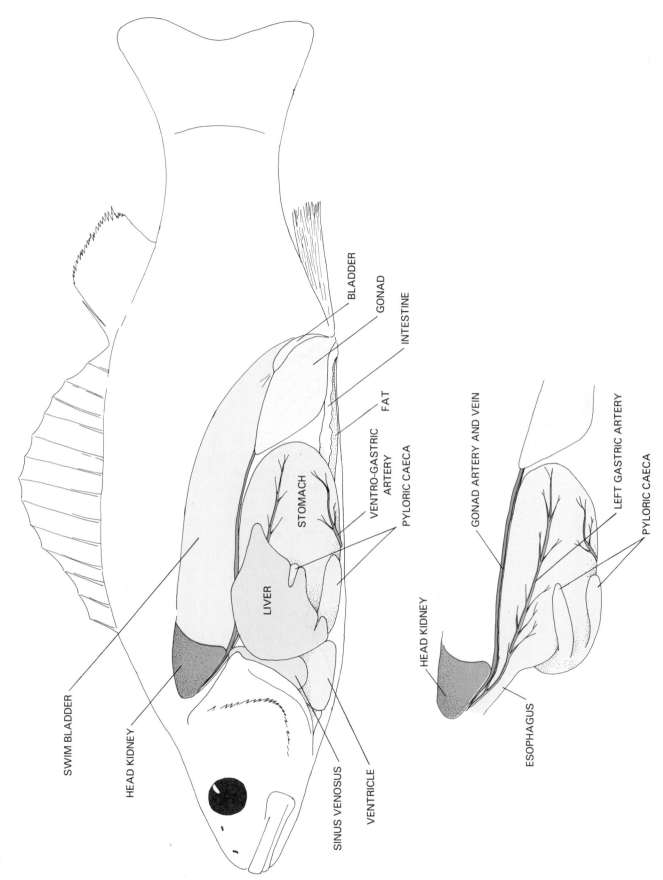

Figure 8.4 Visceral organs and blood vessels on the left side of the perch. In the lower figure the liver is removed to reveal the deeper structures.

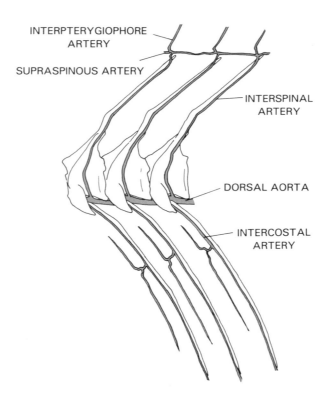

Figure 8.5 Segmental branches of the dorsal aorta.

Branches of the intercostal arteries in the region of the pectoral and pelvic appendages also serve these structures.

4. The *cutaneous artery* (ku-TA-ne-us) (not illustrated) forms as an anastomosis of the superficial branches of the intercostal. The cutaneous artery runs horizontally near the lateral line and serves the superficial layers of the myomeres.

THE SYSTEMIC VEINS

If triple- or double-injected specimens are available, a dissection of the venous circulation may be worthwhile. Attempts to dissect uninjected veins are likely to prove frustrating. In general, the smaller veins accompany the smaller arteries. For example, the intercostal veins are adjacent to the cutaneous arteries. The functional difference between the venous and arterial circulation is that arteries carry blood away from the heart and terminate in a capillary bed while veins return blood to the heart and end in the sinus venosus. Some vessels not only originate but also end in capillary beds (for example, in the kidneys and liver) and are therefore known as portal vessels. Other large veins are double, while their arterial counterpart is single.

Most of the valves controlling the direction of blood flow in the fish circulatory system are found in the heart. Although venous valves have not been reported in bony fishes, some elasmobranchs are known to have valves in the segmental tributary veins at the entrance to the major collecting veins and between the *ductus Cuvieri* and the cardinal sinuses.

A. The Cardinal Venous System
1. The *cranial (anterior) cardinal veins* (KAR-di-nal) (figs. 8.2 and 8.7) drain all the blood from the head into the common cardinal veins. Beside the endocranium, this vessel is known as the *lateral head vein* (not illustrated). The lateral head vein receives four principal tributaries from the head as follows: (a) The *cranial cerebral vein* drains blood from the rostrum and the eye. (b) Two *median cerebral veins* drain blood from the brain case. (c) The *caudal cerebral vein* drains blood from the caudal brain case.
2. The *common cardinal vein* (duct of Cuvier) (figs. 8.1 and 8.7) receives the cranial cardinal vein from in front and the caudal cardinal vein from behind. A small tributary from the lower

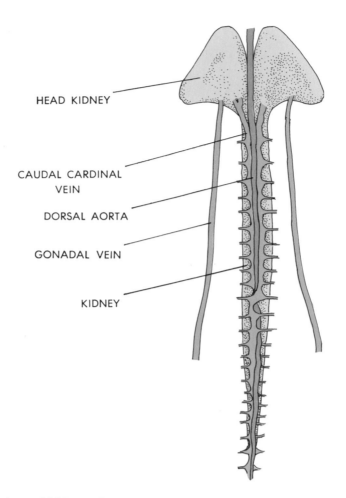

Figure 8.6 Caudal cardinal veins and kidneys. Ventral view.

jaw, sometimes called the *jugular vein* (JUG-u-lar) (not illustrated), also empties into the common cardinal. The two common cardinals drain to the sinus venosus.

3. *Caudal cardinal veins* (fig. 8.6) originate at the caudal end of the trunk cavity from renal and segmental (intercostal and interspinous) veins. The caudal cardinals are found on either side of the dorsal aorta and between the kidneys.

4. The *renal veins* (RE-nal) are short vessels draining the kidneys and opening into the caudal cardinals. Most freshwater teleosts appear to lack a renal portal system—although the perches have some segmental veins from the body that drain into the kidney capillary network thus forming at least a partial renal portal system.

5. *Gonadal veins* from the ovary or testes extend from the dorsal surface of the gonads to the expanded sinuses of the caudal cardinals in the head kidneys.

B. The Hepatic Sinus and Hepatic Portal System

1. The *hepatic veins* (fig. 8.2) receive blood from the liver and open directly into the sinus venosus. The perch has four (2 pairs) hepatic veins. The ventral pair open to the para-atrial chambers of the sinus venosus. The dorsal pair open to the common cardinal veins at their entrance to the sinus venosus.

2. The *hepatic portal vein* (POR-tal) (not illustrated) receives blood from the spleen, intestines, stomach, pancreas, and swim bladder and distributes blood to the sinusoids of the liver.

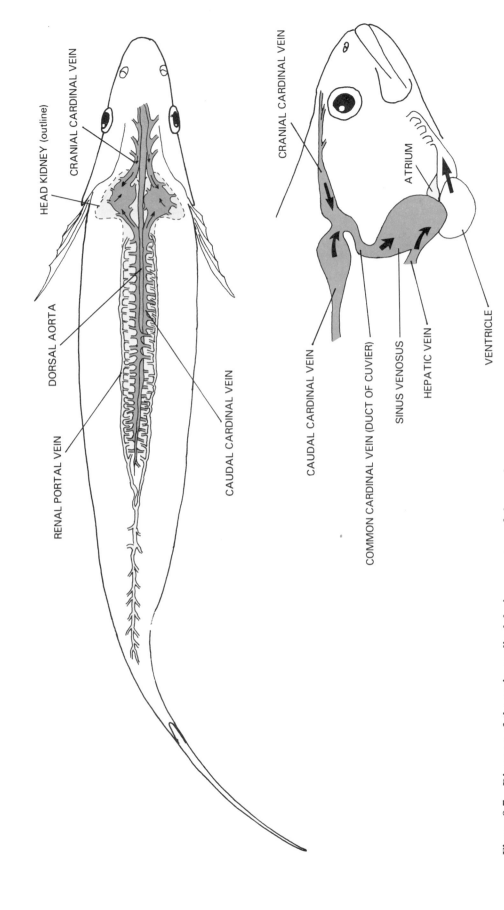

Figure 8.7 Diagram of the main cardinal drainage system of the perch.

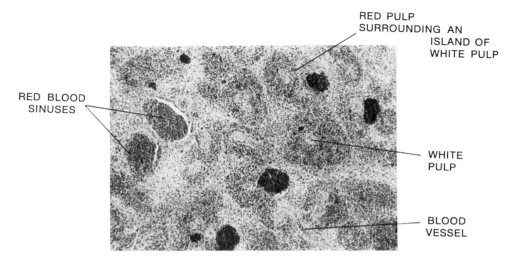

Figure 8.8 Section of the fish spleen × 100.

The hepatic portal system is uninjected so you will not be able to trace these vessels. The following are tributaries of the hepatic portal vessel but are not illustrated: (1) The *splenic vein* drains the spleen and adjacent parts of the stomach. (2) A *gastric vein* drains from the stomach. (3) The *pancreaticoduodenal vein* (pan-kre-at-i-ko-doo-o-DE-nal) drains blood from the capillaries of the pancreas, the pyloric caeca, and from the adjacent duodenum. (4) The *intestinal vein* drains the intestine. (5) Several *cystic* (SIS-tik) and *duodenal veins* drain the capillaries of the gall bladder and adjacent duodenum.

THE SPLEEN

The spleen of fishes (fig. 8.8) is responsible for the production of blood cells in both young and adult animals. There are other blood-forming tissues in the fish such as the kidney, but blood does not form in the fish bone marrow as it does in birds and mammals. The blood cells accumulate in islands or sinuses in the spleen after they are formed. The formation and maturation of the blood cells takes place in the red and white pulp.

The spleen of fishes may have other functions but these have not been documented.

SUGGESTED READINGS

Allis, E. P. 1912. The pseudobranchial and carotid arteries in *Esox, Salmo* and *Gadus,* together with a description of the arteries in the adult *Amia. Anat. Anzeiger* 41(5):113–142.

Cobb, J. L. S. 1981. Gap junctions in the heart of teleost fish. *Cell Tiss. Res.* 154:131–134.

Greco, E. S. and B. Tota. 1981. The ventricular myocardium in fishes, aspects of comparative morphology, physiology and pharmacology. 3. The neurohumoral regulation. *Boll. Soc. Ital. Biol. Sper.* 57(21):2151–2157.

Hibiya, Takashi (Ed.) 1982. *An atlas of fish histology. Normal and pathological features.* Tokyo/ Stuttgart/ New York. Kodansha Ltd.; Gustav Fischer Verlag xii, 147 pages.

Olson, K. R., K. B. Flint, and R. B. Budde. 1981. Vascular corrosion replicas of chemo baro receptors in fish: The carotid labyrinth in Ictaluridae and Claridae. *Cell. Tiss. Res.* 219(3):535–541.

Randall, D. J. 1970. The circulatory system. In *Fish physiology,* ed. W. S. Hoar and D. J. Randall, vol. IV, pp. 133–172. New York: Academic Press.

Santer, R. M. 1985. Morphology and innervation of the fish heart. In *Advances in anatomy, embryology, and cell biology,* vol. 89, pp. 1–102. New York: Springer-Verlag.

Satchell, G. H. 1970. A functional appraisal of the fish heart. *Federation Proceedings* 29(3):1120–1123.

———. 1971. Circulation in fishes. In *Experimental biology,* vol. 18, pp. i–x and 1–131. Cambridge Monographs.

Yasutake, W. T. and J. H. Wales. 1983. Microscopic anatomy of Salmonids: An atlas. *U.S. fish and wildlife service,* Resource Publication 150, Washington, D.C., vi, 189 pages.

Chapter 9
Excretory and Reproductive Systems

The combined *reproductive* (re-pro-DUK-tiv) and *excretory* (EKS-kre-to-re) systems are referred to as the *urogenital system* (u-ro-JEN-i-tal). In the perch, the excretory system is similar in both sexes but the reproductive systems of males and females are quite different.

THE EXCRETORY SYSTEM

The Kidney

Kidneys (KID-nes) are specialized structures that remove the wastes formed by the breakdown of amino acids. The waste product is ammonia, which is very toxic. The toxicity is reduced if the ammonia is converted to some other compound (such as urea or uric acid) or if the ammonia is diluted in sufficient water. The perch kidney removes both dilute ammonia and nontoxic compounds.

The kidney tubules (figs. 9.1 and 9.2) of fishes are contorted ducts with one expanded end (*capsule*) surrounding a ball of blood capillaries—*glomerulus* (glo-MER-u-lus)—and the other end opening to a large drainage duct called the *archinephric duct* (ar-ki-NEF-rik). Waste material in the blood is filtered into the capsule from the glomerulus and passes through the tubules to the archinephric duct and is excreted through the cloaca.

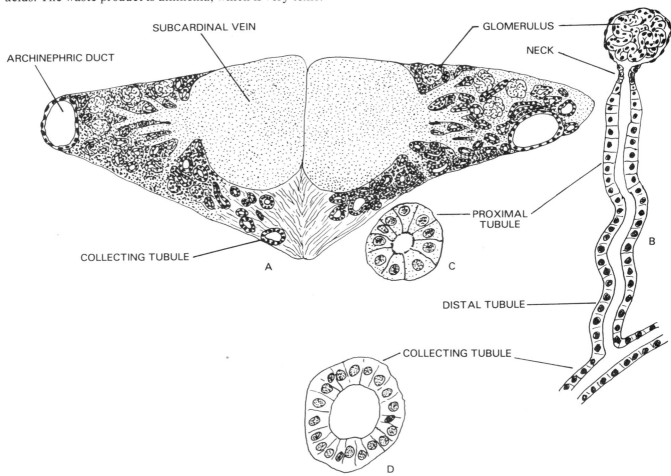

Figure 9.1 Drawing of (A) fish kidney cross-sectioned approximately halfway between anterior and posterior extremes and (B) a diagram of a single fish kidney tubule. Cross sections of proximal (C) and collecting tubules (D) are illustrated below the complete kidney section.

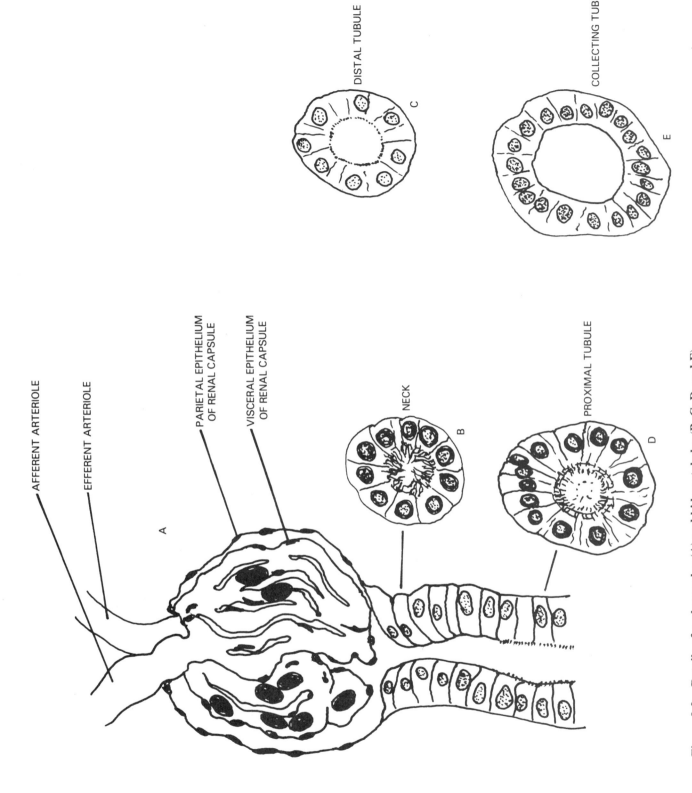

Figure 9.2 Details of a glomerulus (A) and kidney tubules (B, C, D, and E).

Lymphoid cells are located between the tubules and form blood cells in a maturation process similar to the process taking place in the spleen.

The terminology of vertebrate kidney tubules is very complicated because there is one set of terms for the kidney tubules of adult vertebrates based upon a very speculative evolutionary history and another set of terms describing the embryonic stages. Originally, biologists thought that embryonic development repeated the evolutionary stage of that species. Consequently, the embryonic terms were also used for adult kidney tubules. As it became evident that this *biogenetic law* (bi-o-je-NET-ik) was not reliable, a "new" set of terms was devised to reflect the evolutionary sequence of kidney tubules and the established terms were reserved for embryonic stages.

The first kidney tubules to develop embryonically form at the extreme cranial end of the embryonic trunk. Thus, in both time and position, these tubules are indeed "*pronephric*" (pro-NEF-rik). The first few embryonic pronephric tubules form by invagination from the coelomic cavity into the nephrogenic tissue and fuse together at their innermost tips to form the archinephric duct. Once initiated the archinephric duct grows caudally in the nephrogenic tissue with subsequent tubules forming by evagination from the archinephric duct. The tubules formed in this manner develop a direct association with a small knot of blood capillaries (glomeruli) so the tubules no longer need to drain the coelomic fluid to excrete metabolic wastes. Nevertheless, a few of those tubules may also have openings or *nephrostomes* (NEF-ro-stoms) to the coelomic fluid. This second group of embryonic tubules form the *mesonephros* (MES-o-nef-ros). After forming the *mesonephric tubules* the archinephric duct continues to grow caudally and ultimately reaches and opens to the embryonic cloaca.

A hypothesis explaining the origin of vertebrate kidney tubules suggests that the ancestral vertebrate had a single pair of nephric units in each body segment. Each nephric unit was similar to the pronephric tubule with an opening to the coelom and a glomerulus adjacent to the coelom. These segmentally arranged tubules make up a hypothetical *holonephros* (HOL-o-nef-ros).

The most rostral tubules of the holonephros are the *pronephros* of embryonic forms. The holonephric tubules caudal to the pronephros are termed *opisthonephros* (o-PIS-tho-nef-ros).

The opisthonephros is the kidney of the adult fish. This differs from the hypothetical holonephros in two ways; first, the opisthonephros tubules are not segmentally arranged, and second, the pronephric portion is no longer associated but is either degenerate (in females) or forms gonadal ducts (in males).

In many fish the archinephric duct drains both testes and opisthonephros but in the teleost fish (including the perch) the archinephric duct drains only the kidney and a new sperm duct is formed.

The tubules have a capillary bed around them which helps to resorb water from the excretory material in the tubule. This capillary bed is from branches of the renal portal vessel and these empty into vessels that drain into the renal vein. The glomeruli are formed from branches of the renal artery and drain (eventually) into the renal veins.

1. The *opisthonephroi* are paired elongated masses tightly pressed against the dorsal body wall. Often the caudal parts of the two masses are fused in the midline but cranially they are usually separated by the dorsal aorta.

    The cranial end of the opisthonephroi is expanded into a head kidney consisting primarily of blood sinuses. Caudally the kidneys are deflected ventrally on the caudal body wall and end on the caudal surface of the urinary bladder.

2. The *Wolffian ducts (opisthonephric ducts)* (fig. 9.3) are long structures draining the kidney but they may be seen as very short tubules from the caudal terminal end of the kidneys to the male bladder. In the female these ducts open into the urogenital sinus.

3. The *urinary bladder* (U-ri-ner-e BLAD-der) is a small vesicle at the most caudal, ventral area of the body cavity. The bladder is formed from the dorsal wall of the embryonic male cloaca. The cloaca is no longer present in the adult perch. The bladder is incorporated into the oviducts in the female to form a *urogenital sinus*.

4. The *urinary papilla* (pa-PIL-a) is the external projection of the male urinary bladder. Waste materials are extruded through a *urinary pore* of the papilla. In the female the excretory materials are extruded through a urogenital pore.

REPRODUCTIVE SYSTEMS

The gonads of both sexes of vertebrates have a common early development during which the mesenchymal cells of the genital ridge are arranged into *primary sex cords* and *rete cords* (RE-te). These cords will either connect with embryonic kidney tubules and form ducts of the male reproductive system or degenerate to await the formation of *secondary sex cords* that form the lamina of the female ovary. In fish the adult gonads may both be active organs so the animal is hermaphroditic or

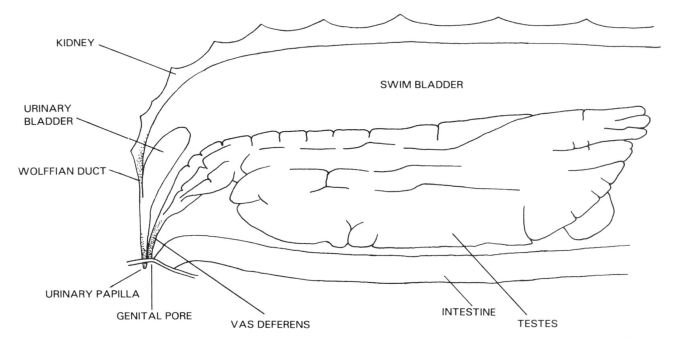

Figure 9.3 Lateral view of the male urogenital organs.

the sexes may be completely separate. In young hermaphroditic forms the testes usually predominates over the ovary but in older fish the ovary predominates.

A. Male Reproductive System
 1. The *testes* (TES-*tez*) (fig. 9.4) are paired, lobulated structures, caudal to the duodenum and stomach, ventral to the air bladder, and dorsal to the intestine. Longitudinal folds run the length of the testes and the vas deferens will be found inside the largest and most central of these folds. *With the help of a dissecting microscope, pick away the tissue of the testes and expose the vas deferens.*
 2. The *vas deferens* (vas DEF-er-enz) (see above) of the two sides join in the caudal midline to form a genital sinus that opens to the exterior between the urinary papilla and the anus. The vas deferens will be located adjacent to the spermatic artery.
 3. The *genital pore* is the external opening of the combined vas deferens.

B. Female Reproductive System
 1. The *ovary* (*O*-va-re) of the female is a single large sack of eggs located in the same position in the trunk cavity as were the testes in the male (fig. 9.5). The single sack probably represents a single ovary and oviduct but there is some evidence the ovaries are fused together.
 2. The *oviduct* (*O*-vi-dukt) in the perch is in two parts. The eggs are actually expelled from the female through an abdominal pore just caudal to the anus. The lining of the pore is folded, internally, into a funnel-like structure that meets the coelomic portion of the oviduct. The coelomic part of the oviduct is formed during embryonic development by a fold of the peritoneum that envelopes the ovary and forms a "tunnel" from the ovary to the abdominal pore.

SUGGESTED READINGS

Barbieri, M. C., G. Barbieri, and M. A. Marins. 1981. Anatomy and histology of the ovary of the fish *Geophagus brasiliensis*. *Rev-Bras-Biol*. Rio de Janeiro, Academia Brasilera de Ciencias 41(1):163–168.

Budd, J. and J. P. Schroeder. 1969. Testicular tumors of yellow perch, *Perca flavescens* (Mitchell). *Bull. Wildl. Dis. Ass.* 5:315–318.

Endo, M. and M. Kimura. 1982. Histological and enzyme histochemical studies on the nephrons of the freshwater fishes *Cyprinus carpio* and *Carassius auratus*. *J. Morphol.* 173(1):29–34.

Grier, H. J., J. R. Linton, J. F. Leatherland, and V. L. de Vlaming. 1980. Structural evidence of four different testicular types in teleost fishes. *Amer. Jour. Anat.* 159(3):331–346.

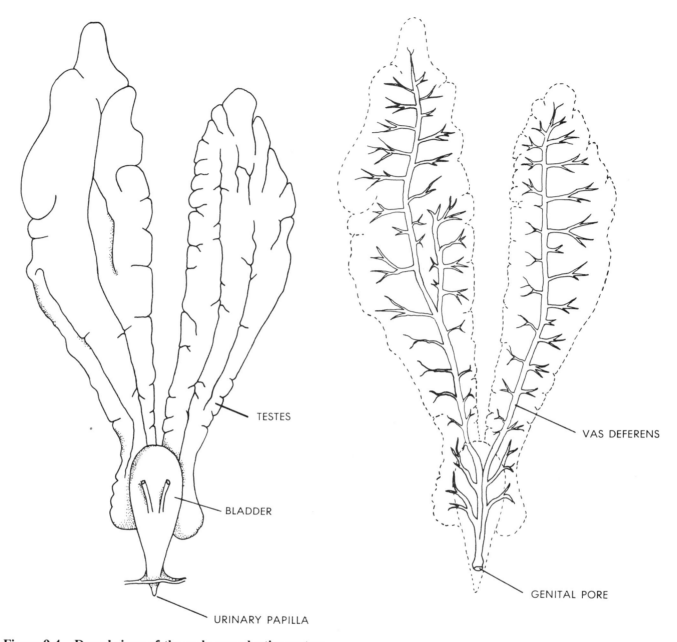

Figure 9.4　Dorsal views of the male reproductive system.

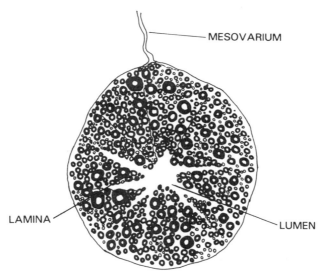

Figure 9.5 Semidiagrammatic cross section of the ovary and associated membranes. Ova erupt from their follicle into the lumen of the ovary and pass through a short oviduct opening between the anus and urinary pore. This arrangement is referred to as a *cystovarian* reproductive system. In a *semicystovarian* system the ova erupt to the body cavity and pass through a funnel-like groove to the genital pore.

Jellyman, D. J. 1980. Age, growth and reproduction of perch, *Perca fluviatilis* in Lake Pounui, New Zealand. *New Zealand Jour. Marine and Freshwater Research* 14(4):391–400.

Oguri, M. 1980. A histological investigation on the juxtaglomerular cell granules in fish kidneys. *Bull. Jpn. Soc. Sci. Fish.* 46(7):798–800.

———. 1980. Presence of juxtaglomerular cells in trout kidneys. *Bull. Jpn. Soc. Sci. Fish.* 46(3):295–298.

———. 1982. Relationship between renal tubule and vascular pole of glomerulus in the holocephalan kidney. *Bull. Jpn. Soc. Sci. Fish.* 48(6):771–774.

Volodin, V. M. 1979. Fertility of the perch, *Perca fluviatilis* from the Rybinsk reservoir, Russian SFSR, USSR. *Vopr. Ikhtiol.* 19(4):672–679.

Zelenkov, V. M. 1981. Early gametogenesis and sex differentiation in the perch, *Perca fluviatilis*. *Vopr. Ikhtiol.* 21(2):331–336.

Chapter 10
The Nervous System

The vertebrate nervous system is customarily arranged into four major divisions: (1) the central nervous system; (2) the general peripheral nervous system; (3) the autonomic nervous system; and (4) the special sense organs. Each of these divisions has additional subdivisions that categorize more specific regions. The central nervous system includes the brain and the spinal cord. The peripheral nervous system consists of cranial and spinal nerves, and branches of the cranial and spinal nerves. The autonomic nervous system is divided into parasympathetic and sympathetic components. Special sense organs of the perch are the lateral line organ, the nose, the ear, and the eye. *Many of the structures of the nervous system are microscopic in size and therefore require histological techniques for their study.* The special sense organs, except the eye and ear, fall in this category as does the internal fine structure of all parts of the nervous system. The major concern of this manual is with the gross structure of the perch and this applies to the nervous system as it does to the other organ systems. In some instances, prepared slides are available from biological supply firms that might be used in conjunction with your dissection.

PREPARATION OF THE CENTRAL NERVOUS SYSTEM FOR OBSERVATION

Remove the skin of the head and with a sharp, strong scalpel carefully slice away the bone of the skull behind the eyes. As the bone of the skull dries it will become more difficult to cut through. This dissection will expose the internal ear. Be careful not to destroy the delicate ear during the dissection. If the ear is to be studied, you should investigate these parts as soon as the ear is exposed and before you uncover the brain.

THE CENTRAL NERVOUS SYSTEM

The Brain (fig. 10.1)

The brain is embedded in a gelatinous mass that must be removed before the brain can be studied. A pigmented membrane, the *meninx* (ME-ninks), covers the brain directly. *This membrane should be removed.*

From rostral to caudal, there are four major subdivisions of the brain: telencephalon, diencephalon, mesencephalon, and rhombencephalon. The specific structures of these subdivisions are as follows:

Telencephalon (tel-en-SEF-a-lon)	Olfactory lobes, cerebral hemispheres
Diencephalon (di-en-SEF-a-lon)	Thalamus, hypothalamus, inferior lobes, saccus vasculosus
Mesencephalon (mez-en-SEF-a-lon)	Optic lobes, torus semicircularis
Rhombencephalon (rom-ben-SEF-a-lon)	
Metencephalon (met-en-SEF-a-lon)	Cerebellum
Myelencephalon (mi-eh-len-SEF-a-lon)	Medulla oblongata

In most instances, those parts of the fish brain that are used the most are largest. For example, bottom-feeding fish are dependent on their sense of smell for locating food and they have substantial olfactory lobes and a large telencephalon. Surface-feeding fish are generally dependent upon sight for locating food and they have large optic lobes and a smaller olfactory apparatus. Thus, it is possible to correlate brain structures with habits and habitats of fish. Observe the following structures with these functional correlates in mind:

Dorsal Aspect of the Brain (fig. 10.2)

1. *Olfactory lobes* (ol-FAK-to-re lobs) are the most rostral portion of the brain. Caudally, the olfactory bulb is distinct from the telencephalon but, rostrally, the bulbs blend imperceptibly into the olfactory nerve.
2. The *telencephalon* is the enlarged forebrain just caudal to the olfactory bulbs. This portion of the fish brain is homologous to the cerebral hemispheres of "higher" vertebrates.
3. The *pineal organ* (PIN-e-al) is a large dark-colored body just to the right of the dorsal midline between the telencephalon and optic lobes.

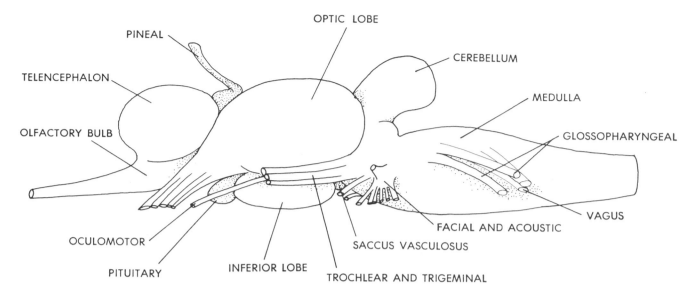

Figure 10.1 Lateral view of the brain.

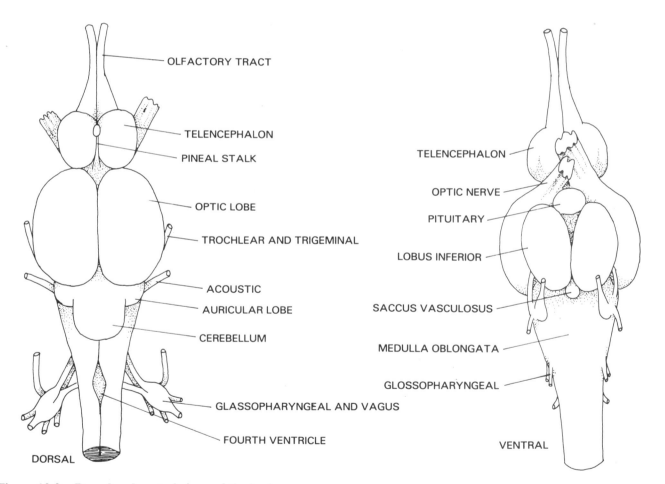

Figure 10.2 Dorsal and ventral views of the brain.

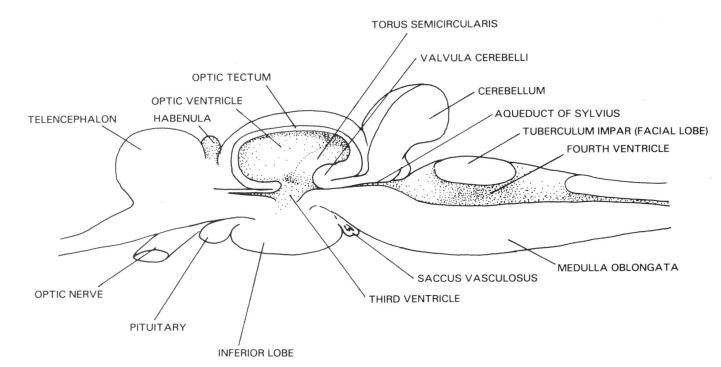

Figure 10.3 Midsagittal view of the brain.

The pineal has a counterpart on the left called the *parapineal* organ. In the perch the parapineal is reduced or absent. In some vertebrates one of these organs (pineal or parapineal) serves as a median photoreceptor. This does not appear to be the functional role of either structure in the perch.

4. The *saccus dorsalis* (SAK-kus dor-SA-lis) is the highly vascularized roof of the diencephalon. In injected specimens, the saccus dorsalis is impregnated with the injection mass which makes it easily distinguishable.

In tetrapods the roof of the diencephalon projects into the third ventricle and is called the tela choroidea.

5. *Optic lobes* (OP-tik), just caudal to the pineal and paraphysis, are the largest structures of the brain. These lobes form the midbrain or mesencephalon.
6. The *cerebellum* (ser-e-BEL-um) is an unpaired, midline structure just caudal to the optic lobes. The caudal end of the cerebellum, the *corpus* (KOR-pus), projects dorsally above the fourth ventricle. Laterally the cerebellum has rounded enlargements, the *auricular lobes* (aw-RIK-u-lar), that project partially into the fourth ventricle. The cranial end of the cerebellum projects forward between the optic lobes as the *valvula cerebelli* (VAL-vu-la). The latter can be seen in a midsagittal section (fig. 10.3).

The corpus is thought to be involved in locomotor activity exclusively. In those fish with electric organs the cerebellum is greatly enlarged. Usually this enlargement is of the valvula but it may also involve the corpus in some species.

The valvula is concerned principally with the lateral line system as are the auricles, but the auricles are additionally concerned with the inner ear (see page 71). Fish that have a large valvula cerebelli also have a better developed lateral line organ. In midsagittal section the auricles can be seen in the caudal lateral corner of the optic ventricle where they are called the *tori semicirculares* (TO-ri sem-e-ser-ku-LAR-is).

7. The *medulla oblongata* (me-DUL-a ob-long-GA-ta) is the most caudal division of the brain. Cranial nerves III to X emerge from this structure and are described in the chart on page 73. The caudal *fourth ventricle* is the open chamber in the dorsal half of the medulla. This chamber is roofed by a tela choroidea, which was probably removed during your dissection.

The lateral cranial walls of the fourth ventricle are expanded into two lobes, the *tuberculi impar* (tu-BER-ku-li IM-par), that meet in the midline at the cranial end of the fourth ventricle.

The medulla oblongata has three important gray matter areas in various fishes. The most cranial areas are continuous with the cerebellum and form the cranial border of the fourth ventricle.

In those fishes in which these areas are well developed, they are designated *somatic sensory lobes* (so-MAT-ik SEN-so-re) and they receive stimuli from tactile (touch) endings of the skin. These lobes are not grossly distinguishable in the perch. The cranial dorsal walls of the fourth ventricle of the perch are enlarged as *facial lobes* (FASH-al) and these two lobes are fused in the dorsal midline as the tuberculum impar. The facial lobes are sensory centers for taste located in the skin. The caudal lateral walls of the medulla contain the *vagal lobes* (VA-gal) that are sensitive to impulses from taste receptors in the mouth and pharynx. The perch has no skin taste buds and consequently very small vagal lobes.

Ventral Aspect of the Brain (fig. 10.2)

Some of the structures described under the dorsal aspect of the brain will also be seen here. Those descriptions will not be repeated.

1. The *optic nerves* will be seen emerging from the cranial ventral corner of the optic lobes. The nerves cross beneath the telencephalon so the nerve from the left optic nerve extends to the right eye and the nerve of the right optic lobe serves the left eye. There is no optic chiasma in the perch.
2. The *pituitary* (pi-too-i-tar-e) or *hypophysis* (hi-POF-i-sis) in the perch is located just caudal to the crossed optic nerves and rostral to the inferior lobes. The pituitary is an endocrine gland that secretes several different hormones.
3. The *lobi inferiores* are paired lobes of the *hypothalamus* (hi-po-THAL-a-mus) (diencephalon) that project ventrally from the base of the brain. These lobes appear to be primarily concerned with sensory perception. Fiber tracts in the inferior lobes also connect with olfactory, gustatory, and optic centers.
4. The *saccus vasculosus* (vas-kyoo-LO-sus) is a single vascular saclike expansion of the ventrocaudal floor of the hypothalamus. It is thought to be sensitive to changes in fluid pressure and therefore an organ for depth perception.

Midsagittal View of the Brain (fig. 10.3)

One of the principal characteristics of the vertebrate central nervous system is that it is hollow. The cavity of the spinal cord is the *central canal* (SEN-tral ka-NAL). In the brain, the central canal is expanded into a series of ventricles; one ventricle is located in each subdivision of the brain. The two lobes of the telencephalon contain the two most rostral ventricles. The diencephalon contains the third ventricle and ventrally the lobi inferiores and saccus vasculosus form visible expansions of the hypothalamus. An optic ventricle forms the interior of each of the mesencephalic lobes. Caudally, a fourth ventricle is located in the dorsal half of the myelencephalon. The paired lateral ventricles are seen if the telencephalon is cross sectioned but cannot be viewed in midsagittal section.

1. The *optic ventricles* are found inside the optic lobes. The superficial tissue of the optic lobes forms the *optic tectum* (TEK-tum).
2. The *valvula cerebelli* is a midline structure projecting forward into the area of the caudal floor between the optic ventricles (See the description of the cerebellum previously discussed.)
3. The *torus semicircularis* is an enlargement in the caudal floor of each optic ventricle. The two tori semicirculari are separated from each other by the valvula cerebelli. The tori semicirculari are homologous to the *caudal corpus quadrigeminum* (kwad-ri-JEM-i-num) (auditory reflex centers) of mammals.
4. The *third ventricle* is a narrow midline channel projecting ventrally from between the two optic ventricles toward the pituitary gland. A narrow passage runs forward from the central chamber to the telencephalic ventricles, and a caudal channel, the *aqueduct of Sylvius* (AK-we-dukt of SIL-ve-us), joins the third ventricle with the fourth.
5. The *infundibulum* (in-fun-DIB-u-lum) is a short stalk running from the floor of the diencephalon to the pituitary. A portion of the third ventricle extends to the center of the infundibulum.
6. The *fourth ventricle* is the cavity of the dorsal half of the myelencephalon.

Cross Section of the Brain

Prepared and specially stained slides will be necessary to see the fiber tracts and nuclear areas of the brain. Some of the major features may be identified from sections made with a razor blade and examined with a dissecting microscope. Figure 10.4 depicts features evident by both methods of preparation. This section was selected to show a large variety of features.

Nerve fibers, *axons* (AX-ons), and *dendrites* (DEN-drits) of most central nervous system nerves are surrounded by a white, fatty substance called *myelin* (mi-e-lin). Cell bodies, *nuclei* (NOO-kle-i), and *synapses* (sin-APS-ez) are not covered by myelin and appear gray. Gray areas in the brain are usually referred to as *nuclei, ganglia* (GANG-gle-a), or when arranged in layers, as *cortex* (KOR-teks), *lamina* (LAM-i-na), or *strata* (STRA-tah). The *gray matter* areas of lower vertebrates are principally concerned with sensory perception and/or the initiation and coordination of motor activity.

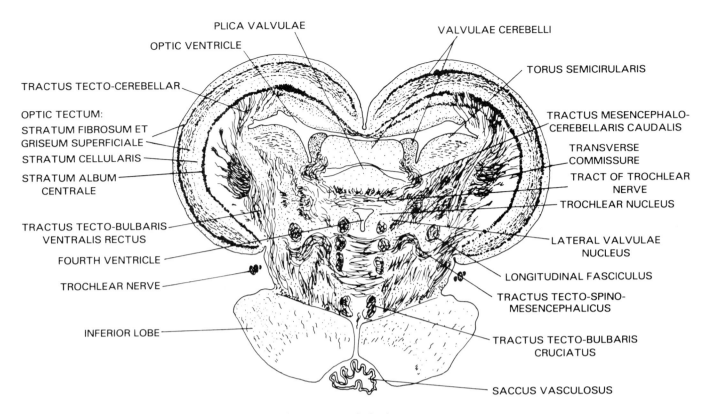

Figure 10.4 Cross section of the perch midbrain (mesencephalon).

Bundles of nerve fibers (*white matter*) are usually referred to as *tracts* (trakts), *fasciculi* (fa-SIK-u-li), or *commissures* (KOM-i-shurs).

The following gray matter areas and fiber tracts are illustrated in figure 10.4.

1. The *optic tectum* is the roof of the optic lobe. The tissue is composed of layers of cells whose nuclei appear in layers or rows. These layers are called strata. There are three major strata in the optic tectum. From superficial to deep they are as follows: (a) *stratum fibrosum et griseum superficialis* (fi-BRO-sum et GRIS-e-um); (b) *stratum cellularis* (sel-u-LAR-is); and (c) *stratum album centrale* (AL-bum sen-TRA-le). This tissue is both a reflex and coordinating center.
2. The *Torus semicircularis* is to the ears what the optic lobes are to the eyes. This region also includes the sense of equilibrium, also a function of the ear.
3. The *Valvula cerebelli* is a portion of the cerebellum (see previously discussed section on the cerebellum) projecting forward under and between the optic lobes. The gray matter of this structure consists of a central molecular portion and two lateral granular laminae. Part of the valvula cerebelli is folded beneath the molecular lamina as the *plica valvula* (PLI-ka). *Lateral valvula nuclei* of the valvula cerebelli are located below the granular laminae.
4. The *trochlear nuclei* (TROK-le-ar) are located medial to the lateral nuclei of the valvulae.
5. *Tracts* (fig. 10.4), passing from rostral to caudal, are cut in cross section and those running vertically or transversely are cut lengthwise in a transverse section. In general, fiber tracts connect nuclear areas with each other or with sensory or motor organs. The names of fiber tracts usually contain the areas they connect.
    a. *Tractus tecto-bulbaris ventralis rectus* connects the optic tectum (tecto) and medulla oblongata (bulbar) by a straight (rectus) ventral fiber tract.
    b. *Tractus tecto-bulbaris cruciatus* (KROO-she-at-us) also connects the optic tectum and medulla but by fibers that cross each other (cruciatus) in the midline.
    c. *Tractus tecto-cerebellar* connects the optic tectum and cerebellum.
    d. *Tractus tecto-spino-mesencephalicus* (mez-en-sef-a-li-cus) connects the optic tectum, spinal cord, and other mesencephalic centers.
    e. *Tractus mesencephalo-cerebellaris caudalis* connects the valvulae cerebelli with the mesencephalon.

CHART OF THE CRANIAL NERVES

NAME	ORIGIN ON THE BRAIN	FUNCTION	DISTRIBUTION
O. Nervus terminalis (ter-mi-NAL-is)	Ventral border of olfactory bulb.	Sensory	Sense endings of snout and olfactory epithelium.
I. Olfactory (ol-FAK-to-re)	Cranial end of olfactory bulb.	Sensory	Nasal epithelium.
II. Optic (OP-tik)	Optic lobes, ventral to telencephalon.	Sensory	Retina of eye.
III. Oculomotor (ok-u-lo-MO-tor)	Caudal ventral end of medulla, between optic lobes and cerebellum.	Motor	Musculus muscularis superior, inferior and medial rectus and inferior oblique.
IV. Trochlear (TROK-le-ar)	Dorsolateral cranial surface of medulla. Extremely small.	Motor	Superior oblique muscle.
V. Trigeminal (tri-JEM-i-nal)	Cranial lateral border of medulla oblongata.	M and S	Jaw muscles, touch endings in skin of the head.
VI. Abducens (ab-DOO-senz)	Cranial ventral end of medulla oblongata.	Motor	Lateral rectus muscle.
VII. Facial portion Acousticofacialis (a-coos-ti-ko-fa-se-AL-is)	Lateral border of medulla oblongata just caudal to trigeminal and with acoustic.	M and S	Skin of head, taste endings.
VIII. Acoustic portion Acousticofacialis	With facial from lateral border of medulla oblongata.	Sensory	Inner ear and lateral line organ.
IX. Glossopharyngeal (glos-o-fah-RIN-je-al)	Caudal lateral border of medulla with vagus.	M and S	Gill muscles.
X. Vagus[1]	Caudal lateral border of medulla with glossopharyngeal.	M and S	Gills, heart, and cranial portion of alimentary tract.

[1]This and the glossopharyngeal are inseparable medial to the lateral ganglion.

f. *Longitudinal fasciculi* are thick tracts which serve many areas of the brain.

g. The *transverse commissure* carries fibers between the right and left sides of the brain.

h. *Trochlear nerve tracts* have connections with the valvulae cerebelli, the optic tectum, and the nerve of the opposite side.

THE CENTRAL NERVOUS SYSTEM

The Spinal Cord

The spinal cord of the perch requires histological techniques for its proper study. Even what might be referred to as a "gross" dissection will require a dissecting microscope.

The major features of the typical vertebrate spinal cord are (1) a central spinal nerve cord extending caudally from the brain and passing through the vertebral neural arches, and (2) a series of paired spinal nerves emerging from the spinal cord and serving various somatic structures. In the perch, the paired spinal nerves pass through vertebral foramina as they emerge from the spinal cord. As a result of this arrangement the nerves are destroyed as the vertebrae are removed. A study of these nerves is therefore extremely difficult. Figure 10.5 illustrates the typical spinal nerves and indicates their relationship to the vertebral column. Figure 10.6 elucidates the arrangement of the first three spinal nerves and a typical spinal nerve of the trunk.

The first three spinal nerves innervate the muscles of the pectoral and pelvic girdles. These nerves form a *plexus* (PLEK-sus) of sorts in which parts of the first and third nerves join before sending branches to the pectoral muscles. The second nerve sends branches to both pectoral and pelvic muscles but does join the other two nerves. In the perch, all three of the spinal nerves concerned in the pectoral-pelvic plexus appear to lack dorsal roots. In addition to serving the fin muscles, the first spinal nerve has an *occipital branch* (ok-SIP-i-tal) extending to the back of the cranium and a *ventral-rostral branch* reaching the hyoid muscles.

The Autonomic Nervous System

Teleosts are said to be the only pretetrapod group of vertebrates to possess a true ganglionated *sympathetic* (sim-pa-THET-ik) *nerve trunk* (Kappers, Huber, and Crosby, 1936. *Comparative Anatomy of the Nervous System of Vertebrates,* Macmillan Co.).

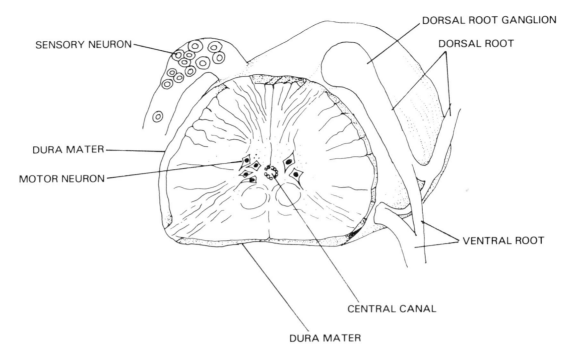

Figure 10.5 **Spinal cord and the arrangement of spinal nerves.**

The ganglia of this trunk are connected with the spinal nerves by *rami communicantes* (RA-mi ka-mu-ne-KAN-tez) after the dorsal and ventral roots of the nerve have joined. Both *afferent* (AF-er-ent) and *efferent* (EF-er-ent) *sympathetic fibers* pass through the *ventral root*. The ganglionated trunk is found just medial to the ribs, within the body cavity.

The *vagus nerve* (a cranial nerve) represents the major parasympathetic portion of the autonomic nervous system. The vagus is described with the other cranial nerves (see the chart of cranial nerves, page 73).

THE SPECIAL SENSE ORGANS

The Lateral Line Organ

The *lateral line organs* of the perch consist of a series of interconnected tubules in the dermis of the skin. One long tubule runs the length of the body on either side, just lateral to the juncture of epaxial and hypaxial muscles. It is these portions that give the organ its name.

In the head, the lateral line structures run through passageways in the dermal bone of the skull. These passageways can be found in the bones around the orbit and in the lower jaw as well as on the top of the skull.

As the tubule passes through a scale (or scale bone) a branch of the tubule opens to the surface. In the horizontal portion of the tubule between the lateral branches are groups of ciliated cells called *neuromasts* (NU-ro-masts) (fig. 10.7). The ciliated end of the neuromast cells project into the lateral line canal and the opposite end is innervated by branches of the *vagus* (lateral line branch) on the trunk and by the *facial* and *glossopharyngeal* nerves on the head.

Water movement in the canal moves the kinocilium of one neuromast away from the stereocilia thus inhibiting the neural activity of the cell. Neuromasts are linked to other neuromasts with the kinocilium at the opposite position on the cell. Thus by the same movement that moves the kinocilium of one neuromast toward the stereocilia, the kinocilium of the linked neuromast is moved away from its stereocilia. Moving the kinocilium toward the stereocilia stimulates the cell and moving the kinocilium away from the stereocilia inhibits activity. Stimulation of one cell and inhibition of the other indicates the direction of movement to the perch.

Experimental evidence also shows that the lateral line organ is sensitive to pressure (acoustic waves) and temperature changes. Several other, as yet unsubstantiated, functions have been claimed for this organ.

The Ear

The *internal ear* of vertebrates is thought to be a specialized portion of the lateral line system of the head. Both the ear and lateral line develop in similar ways; both have similar sensory organs (neuromasts); and both are innervated by nerves from related brain centers.

If the internal ear is to be dissected, the student should also dissect the suborbital portion of the cranial lateral line system and compare the canals.

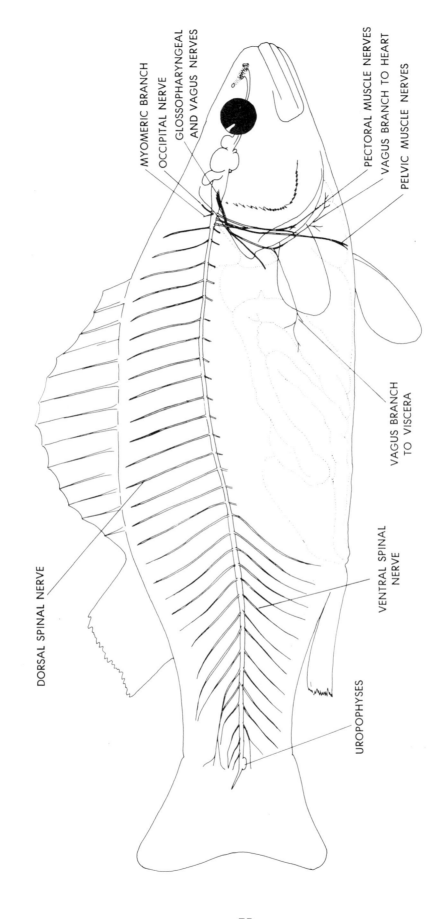

Figure 10.6 The central nervous system and appendicular plexus.

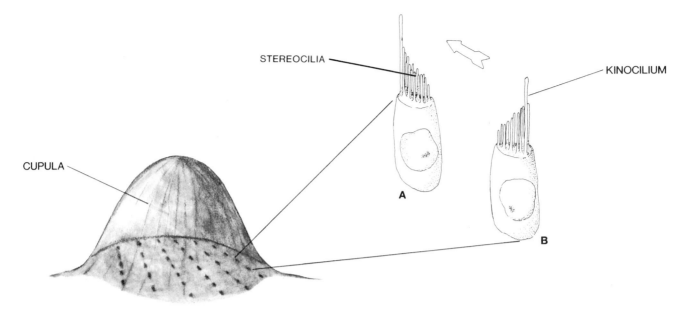

Figure 10.7 Diagram of a lateral line organ. Water movement in the direction of the arrow moves the kinocilium of A away from the stereocilia, thus inhibiting the neural activity of the cell. By the same movement the kinocilium of B is moved toward the stereocilia, thus stimulating cell B. Stimulation of one cell and inhibition of the other indicates the direction of movement of the perch.

Although hearing has been reliably demonstrated in *Perca fluviatilis*, the receptive organ has not been satisfactorily identified. For some fishes it has been postulated that the *saccule* (SAK-*ul*) is the sound receptor area; in others the *utricle* (*U*-tre-kl) has been implicated. The most reliable investigations on this subject have revealed the saccule as the most likely acoustic detector (in goldfish).

It is thought that the swim bladder, working in conjunction with the inner ear, serves as an aid to hearing but fish (sharks) without swim bladders are also capable of hearing. Fish with swim bladders have the lowest auditory thresholds and the highest frequency limits.

The lateral line system of fishes operates as a low-frequency acoustic receptor detecting water displacements when the fish is very close to the source of the displacement (sound). The ear (saccule) detects pressure waves (sound) at a much greater distance from the source of the sound. Both act as acoustic sense organs.

The ear of the perch consists entirely of the internal ear. There is no external or middle ear in fishes. Portions of the ear are inside the cranial cavity (rostral vertical semicircular canal, utricle, and saccule) and part (horizontal and caudal vertical semicircular canals) is embedded in the cartilaginous (prootic, opisthotic, and epiotic) skeleton (see chapter 3).

Preparation of the Inner Ear for Observation. The rostral vertical semicircular canal will be seen just lateral to the brain (see fig. 10.8). With a sharp scalpel, carefully shave away the dermal bone just lateral to the ear. The cartilage can be removed with a pair of fine-pointed forceps under a dissecting microscope.

1. The *semicircular canals* (sem-*e*-SER-k*u*-ka-NALS) are arranged in three planes on each side of the head. Figure 10.8 illustrates these planes as well as the relationship of the semicircular canals to the brain. The names of the semicircular canals indicate their position; the cranial vertical, caudal vertical, and horizontal. At one base of each canal is an enlargement called an *ampulla* (am-PUL-la). Neuromasts (see lateral line previously discussed) are located inside the ampulla. Branches of the *acoustic* nerve serve the neuromasts.

2. The *utricle* is a "football-shaped" enlargement just beneath and slightly caudal to the ampullae of the rostral vertical and horizontal semicircular canals. The wall of the utricle also contains neuromasts served by the acoustic nerve. A caudal extension from the utricle connects the caudal portions of the semicircular canals.

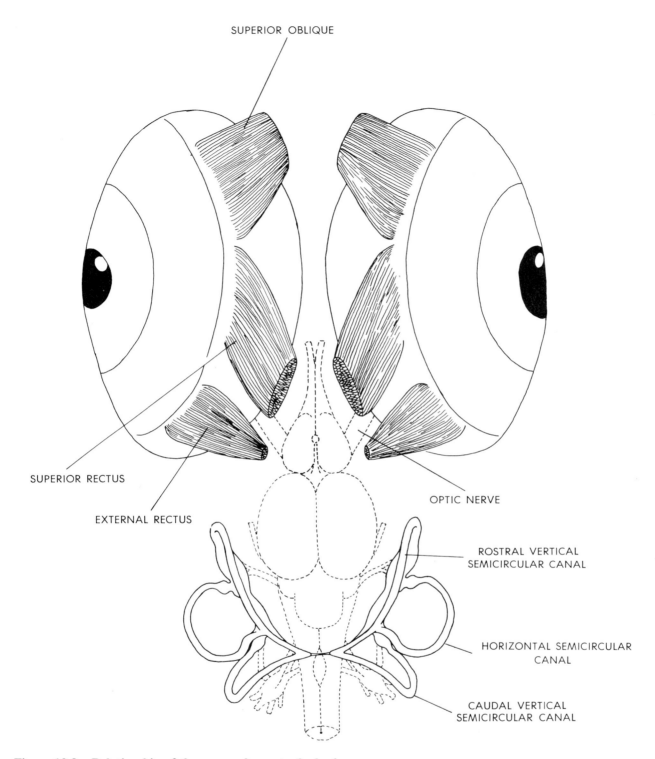

Figure 10.8 Relationship of the eyes and ears to the brain.

3. The *saccule* is a large chamber beneath and caudal to the utricle. The membranous wall of the saccule is extremely thin and easily destroyed. A large "bone," the otolith or sagitta, will be found inside the saccule.

4. The *sagitta* (SAJ-ih-tah) is a relatively large, flat "bone" inside the saccule. Ventrally, the sagitta has a projection that extends into the lagena. Movements of the sagitta activate the neuromasts of the saccule.

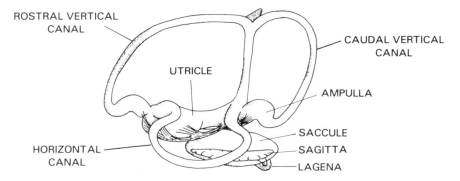

Figure 10.9 Lateral view of the inner ear.

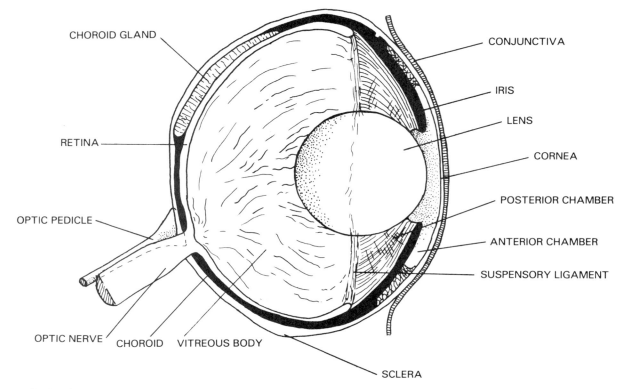

Figure 10.10 Sagittal view of the eye.

5. The *lagena* (lah-JEH-nah) is a small outpocketing from the caudal ventral portion of the saccule. This structure is thought to be homologous to the proximal end of the tetrapod *cochlea* (KOK-le-ah).

The Eye

The structure of the eye of fishes is extremely variable. The description used here will apply only to the Yellow Perch, *Perca flavescens,* and to no other species. For a generalized account of fish eyes, see Walls (1942).

Preparation of the Eye for Observation. If you did not remove the eye from the orbit while studying the muscles of the head, do so now. Using a blunt probe, carefully free the conjunctiva and fascia from the edge of the eye. Cut the extrinsic muscles and the optic nerve so that the longest possible fragments remain attached to the eye and lift the eye from the orbit.

After you have identified the optic nerve and the extrinsic muscles, use a sharp scalpel or razor blade to make a sagittal section of the eye through the midline of the pupil (fig. 10.10).

1. The eye muscles are responsible for moving the eyeball within the orbit. These include the *inferior, superior, lateral,* and *medial recti* muscles and the *superior* and *inferior oblique muscles* (see fig. 4.11). The innervations of these muscles are indicated in the "Chart of the Cranial Nerves" in this chapter.

2. The *optic pedicle* (PED-i-kl) is a flexible cartilage rod adjacent to the optic nerve and holding the eyeball away from the cranium.
3. The *sclera* (SKLE-ra) is the tough outer coat of the eyeball. Notice that some areas of the sclera are thicker than others.
4. The *cornea* (KOR-ne-a) is the thin, transparent, anterior continuation of the sclera.
5. The *choroid layer* (KO-royd) is the black pigmented tissue between the sclera and the retina of the internal half of the eyeball. The iris is the anterior continuation of the choroid layer. The perch has a grayish brown *choroid gland* on the dorsal half of the eyeball in the choroid layer.
6. The *iris* (I-ris) is the heavily pigmented continuation of the choroid layer just internal to the cornea. Fibers connect the iris with the anterior cornea.
7. The *pupil* (PYOO-pil) is the large opening in the center of the iris.
8. The *anterior chamber* (CHAM-ber) of the fish eye is a very small area between the iris and the cornea.
9. The *posterior chamber* of the eye is found between the iris and the lens. The anterior and posterior chambers are connected by the pupil and collectively form the *anterior cavity*.
10. The *lens* (lenz) of the perch is a spherical body, suspended from the anterior edge of the retina by a *suspensory ligament* (sus-PEN-so-re LIG-ament).
11. The *retina* (RET-i-na) is a grayish tissue lining the interior of the posterior half of the eyeball. It is the location of the photoreceptive cells and posteriorly, the retina is continuous with the *optic nerve*. The thickness of the retina is very uneven.
12. The *vitreous body* (VIT-re-us BOD-e) is the viscous fluid filling the *posterior cavity* found between the lens and the retina.

The Olfactory Sense

There are four *nasal apertures* opening into two (right and left) *olfactory sacs* or *chambers*. Water enters the rostral nasal aperture, passes over the *olfactory epithelium* lining the nasal sacs, and passes out the caudal nasal apertures. This movement of water is essential for odor perception and the significance of olfaction is directly related to the strength and evenness of flow of the water current. Thus, fishes living in swift-moving rivers and streams have a good olfactory sense that they use in feeding. Fish living in ponds and lakes with little or no currents have a poor olfactory sense and are dependent upon vision and taste for locating food. Although the Yellow Perch has not been studied in this regard, its close relative, the European Perch (*Perca fluviatilis*), is known to have an olfactory epithelium with 87,000 sensory cells/mm^2. The oval-shaped olfactory chamber has 13 to 18 folds of the epithelium in its anterior half and this epithelium increases with age.

SUGGESTED READINGS

Bone, Q. and R. D. Ono. 1982. Systematic implications of innervation patterns in teleost myotomes. *Brevoria* 470:1–23.

Copland, D. E. 1982. The anatomy and fine structure of the eye in fish. Ciliary type tissue in nine species of teleosts. *Biol. Bull.* 163(1):131–143.

Demske, L. S., D. H. Bauer, and J. W. Gerald. 1975. Sperm release evoked by electrical stimulation of the fish brain: A functional-anatomical study. *J. Expt. Zool.* 191(2):215–232.

Gillian, L. A. 1967. A comparative study of the extrinsic and intrinsic arterial blood supply to brains of submammalian vertebrates. *Jour. Comp. Neur.* 130(3):175–196.

Guma, A. S. R. 1982. Retinal development and retinomotor responses in perch, *Perca fluviatilis*. *J. Fish. Biol.* 20(5):611–618.

Kerr, T. 1942. On the pituitary of the perch (*Perca fluviatilis*). *Quart. J. M. Sci.* 83:299–316.

———.1943. Evolution of the pituitary, with special reference to the teleosts. *Proc. Leeds Phil. Soc.* 4:75–83.

Kleerekoper, H. 1969. *Olfaction in fishes*. Indiana University Press, Boomington, Ind.

McGlone, F. P. and D. H. Paul. 1980. Morphology and electrophysiology of the cerebellum of the perch, *Perca fluviatilis*. *J. Physiology* 303(O):23.

Nieuwenhuys, R. 1962. Trends in the evolution of the Actinopterygian forebrain. *J. Morph.* 111(1):69–88.

Northcutt, R. G. 1978. Fish and amphibian auditory neuroanatomy. *J. Acoust. Soc. Amer.* 64(suppl 1):S 2.

Otten, E. 1981. Vision during growth of a generalized haplochromis species, *Haplochromin elegans*, Pisces, Cichlidae. *Neth. J. Zool.* 31(4):650–700.

Pevzner, R. A. 1981. Fine structure of taste buds of Ganoid teleosts 1. Adult Acipenseridae. *Tsitologiya* 23(7):760–766.

Popper, A. N. and R. G. Northcutt. 1983. Structure and innervation of the inner ear of the Bowfin *Amia calva*. *J. Comp. Neurol.* 213(3):279–286.

Popper, A. N. and W. N. Tavolga. 1981. Structure and function of the ear in the marine catfish, *Arius felis*. *J. Comp. Physiol. and Sensory Neural Behavioral Physiology* 144(1):27–34.

Popper, A. N. and B. Hoxter. 1981. The fine structure of the sacculus and lagena of a teleost fish *Trichogaster trichopterus*. *Hear. Res.* 5(2–3):245–264.

Popper, A. N. 1980. Scanning electron microscopic study of the sacculus and lagena in several deep sea fishes. *Amer. J. Anat.* 157(2):115–136.

Sivak, J. G. 1979. Accommodative lens movements and pupil shape in teleost fishes. *Israeli J. Zool.* 284:218–223.

Tavolga, W. N. 1971. Sound production and detection. In *Fish Physiology*, eds. W. S. Hoar and D. J. Randall, Vol. V, pp. 135–206. New York: Academic Press.

Walls, G. L. 1942. The vertebrate eye and its adaptive radiation. *Cranbrook institute of science bulletin 19*. Bloomfield Hills, Michigan. (Reprinted 1963, Hafner Publishing Company, New York).

Chapter 11
The Endocrine System

INTRODUCTION

The *endocrine system* (EN-do-krin) system is an accumulation of organs and tissues secreting *hormones* (HOR-mons) directly into the bloodstream. Hormones may be polypeptides or steroids and these reach the *target organ* (the organ affected) through the general circulation. It is at the target organ that the hormone exerts the desired effect. Generally a hormone is mediated by some type of receptor at the target organ.

Fishes have been found to have many of the hormones present in the other vertebrates. The organs and tissues producing and secreting the hormones are often very diffuse rather than forming organized glands as is the case in mammals and birds. The functions of the various hormones are, however, often similar. As well, there are at least two additional endocrine organs that are not currently recognized in other vertebrates—the caudal neurosecretory organ and the Stannius corpuscles.

Endocrine glands do not form a distinct anatomical system but instead these glands are often part of another anatomical system and are consequently described with their parent system. For instance, the ovaries and testes are described with the reproductive system although they contain hormone-producing cells. Likewise, the nervous system has many endocrine-secreting neurons and the digestive tract has hormone secretions from the gastric mucosa and from islet cells in the pancreas.

Except for the pituitary it is very difficult to observe these glands at the macroscopic level. Figure 11.1 illustrates the approximate locations of the components of the endocrine system.

THE ENDOCRINE ORGANS

The Hypothalamus

The *hypothalamus* (hi-po-THAL-a-mus) is the floor of the diencephalon. This region consists of the *hypothalamic nuclei* which integrate the central nervous system with endocrine activity and, thus, to somatic functions. These nuclei produce *releasing factors* (hormones) that control synthesis and release of hormones from the pituitary.

The Pituitary

The *pituitary* (pi-TYOO-i-tar-e) or hypophysis (hi-POF-i-sis) is a ventral appendage of the hypothalamic diencephalon. Embryologically, this organ has both neural and epidermal origins. The neural components are from the diencephalon and the epidermal contribution is from the oral cavity of the embryo. In some fish species there is a stalk or *infundibulum* (in-fun-DIB-u-lum) connecting the pituitary to the hypothalamus. Blood vessels and nerve axons emanating from the hypothalamus pass to the pituitary along this pathway. The vessels (called the *hypophyseal portal system*) carry the releasing hormones directly from the nuclei to the cells of the pituitary.

The pituitary of the perch is subdivided into a *neurohypophysis* (the neural component) and releases neurosecretions passed through neuronal axons originating in the hypothalamus (fig. 11.1). The two hormones released here are *oxytocin* (ok-se-TO-sin), involved in reproduction, and *vasotocin* (VAS-o-to-sin) which is osmoregulatory.

The *adenohypophysis* (a-den-o-hi-POF-i-sis), the epidermal division of the pituitary, is further separated into the *pars distalis* (parz DIS-ta-lis) and the *pars intermedia* (in-ter-ME-de-a). These areas are functionally similar to those in birds and mammals. The actual shape of the gland varies considerably from species to species just as it does in other classes of vertebrates. However, molecular structures of the secretions are often identical between species or differ by one or two amino acids in what are often quite large polypeptides.

The following are secretions of the pars distalis:

HORMONE	FUNCTION
prolactin (pro-LAK-tin)	sodium regulation
growth hormone	bone growth
corticotropin (kor-ti-ko-TRO-pin)	regulates steroid secretion from interrenal tissues
gonadotropin (gon-a-do-TRO-pin)	regulates reproduction
thyrotropin (thi-ro-TRO-pin)	regulates thyroid

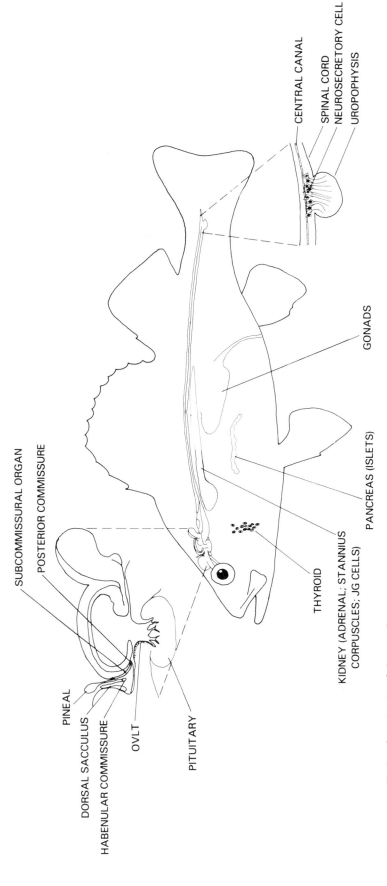

Figure 11.1 Endocrine glands of the perch.

81

The only hormone of the pars intermedia is *melanotropin* (mel-a-no-TRO-pin) which is involved in control of the cutaneous melanophores.

The Organon Vasculosum Laminae Terminalis (OVLT)

The *OVLT* is located in the vicinity of the third ventricle rostral to the infundibulum. In *Perca fluviatilis* the OVLT extends into the third ventricle as a crest. While an endocrine function is suspected for this organ it is known only that it acts as a shunt between the blood and the cerebrospinal fluid (CSF). This structure has been described in other fishes as well as amphibians, reptiles, birds, and mammals.

The Pineal

In the roof of the diencephalon lies the *pineal gland* (PIN-e-al) or *epiphysis* (e-PIF-e-sis). It is considered to be a gland rather than a photoreceptor because it is extensively vascularized. The pineal produces the hormone *melanin* (MEL-a-nin) which is probably involved in control of reproduction. Pinealectomy in fish leads to accelerated gonadal development while exogenous melanin inhibits gonadal growth.

The structure of this gland varies between species but in general consists of the following parts. The *pineal stalk* extends dorsally just anterior to the optic lobe (tectum) and this is expanded into a *pineal vesicle* with a central lumen just beneath the roof of the cranium. The ventral base of the stalk has a *posterior commissure* and a *subcommissural organ* on the posterior portion of the stalk and just anterior to the stalk is a membranous sack, the *dorsal sacculus*. Between the base of the dorsal sacculus and the pineal stalk is a *habenular commissure*. The tissues of the pineal vesicle consist of photoreceptive, supporting, and ganglion cells.

The Thyroid

Although a very diffuse organ arranged as follicles, the *thyroid gland* (THI-royd) of fish is usually associated with the heart or arteries of the head region (ventral aorta, branchial arteries). Occasionally the follicles are further scattered to the head kidneys, eye, brain, esophagus, and spleen. Only sharks have an organized gland similar to mammals. The thyroid hormones are similar in function to those of other vertebrates and target the general body cells to increase oxygen consumption and carbohydrate metabolism. These hormones are also important in body growth and development of the central nervous system.

The Ultimobranchial Gland

The *ultimobranchial gland* (ul-ti-ma-BRANG-ke-al) consists of small nodules of tissue on the dorsal pericardial membrane near the sinus venosus and basibranchial. *Calcitonin* (kal-si-TO-nin), also produced in the thyroid of mammals, has been isolated from this gland. Calcitonin reduces levels of calcium in the blood by increasing bone formation.

The Endocrine Cells of the Pancreas

These cells (arranged as *islets of Langerhans* [LAHNG-er-hanz]) are associated with the pancreas and produce *insulin* (IN-su-lin). This polypeptide is important in the regulation of metabolism and the control of circulating plasma glucose levels.

The Endocrine Cells of the Intestinal Mucosa

Located in the mucosa of the small intestine are specialized endocrine cells producing the hormone *secretin* (se-KRE-tin). This hormone regulates the production of pancreatic digestive enzymes.

The Adrenal

Chromaffin Tissue   These tissues are similar to the adrenal medulla of mammals and are found mixed with interrenal tissues scattered along the head kidney adjacent to the postcardinals. The *chromaffin tissue* (kro-MAF-in) produces adrenalin that alters the activity of visceral organs in the same fashion as does the sympathetic nervous system (for example, increased heart rate and blood pressure).

Interrenal Tissue   *Interrenal tissues* (in-ter-RE-nal) produce steroids such as *cortisol* (KOR-ti-sol), *corticosterone* (kor-ti-KOS-ter-on), and *cortisone* (KOR-ti-son) similar to the mammalian adrenal cortex. Cortisone is the most common adrenal steroid in bony fish and is an osmoregulator targeting the kidney, gill, and gastrointestinal tract.

Corpuscles of Stannius

Secretion of the blood calcium-lowering hormone *hypocalcin* (hi-po-kal-cin) is the function of the *Stannius corpuscles* (STAN-e-us). These are found either on the surface of or embedded in the opisthonephric kidney of teleost fishes (not elasmobranchs).

The Kidney

The hormone *renin* (RE-nin) is produced by juxtaglomerular cells of the teleost. In mammals, this hormone promotes the conversion of angiotensinogen to angiotensin I and this in turn to angiotensin II. Angiotensin II has a steroidogenic action in teleosts and may act directly to increase blood pressure.

The Gonads

As in other vertebrates the *gonads* (GO-nads) produce steroids that in the female *(estrogens)* promote receptivity to the male and encourage the development of secondary sexual characteristics. In the male the sex steroids *(androgens)* encourage male behavioral patterns including territoriality, courtship, and sexual behavior. The androgens may be produced from *Leydig cells* (L*I*-dig) which have been reportedly found between the seminiferous lobules that make up the bulk of the testes. There is a disagreement in that some researchers report that *Sertoli cells* (ser-T*O*-le) or other cells may be responsible for androgen production in fish.

The Caudal Neurosecretory System

Also known as the *urophysial gland* (u-ro-F*IZ*-e-al), the caudal neurosecretory system is, at this time, not well understood. Extracts of the gland have a diuretic effect in mammals and there is other pharmacological and physiological evidence that this secretion acts as an osmoregulatory hormone. The "gland" occurs as an enlargement on the ventral surface of the caudal spinal cord and is situated in a fossa of the urostyle. The fossa contains large neurosecretory cells (*Dahlgren cells* [DAHL-gren]) that secrete into the renal portal system.

SUGGESTED READINGS

American Society of Zoologists. 1983. Symposium: Evolution of endocrine systems in lower vertebrates. *Am. Zool.* 23(3):593–748.

American Society of Zoologists. 1988. Symposium: Comparative endocrinology of the thyroid. *Am. Zool.* 28(2):293–440.

Ball, J. N. and B. I. Baker. 1969. The pituitary gland: Anatomy and histophysiology. In *Fish physiology,* eds. W. S. Hoar and D. J. Randall, vol. II, pp. 1–110. New York: Academic Press.

Benjamin, M. 1982. The ACTH cells in the pituitary gland of the nine-spined stickleback *Pungitius pungitius. Jour. Anat.* 135(1):155–164.

Bulger, E. B. and B. F. Trump. 1969. Ultrastructure of granulated arteriolar cells (juxtaglomerular cells) in kidney of a fresh and a salt water teleost. *Amer. J. Anat.* 124(1):77–88.

Fridberg, G. and H. A. Bern. 1968. The urophysis and the caudal neurosecretory system of fishes. *Biol. Rev.* 43:175–199.

Gorbman, A., W. W. Dickhoff, S. R. Vigna, N. B. Clark, and C. L. Ralph. 1983. *Comparative endocrinology.* New York: John Wiley & Sons.

Ichikawa, T. 1980. Antidiuretic activity of teleost urophysial extracts in the rat. *J. Comp. Physiol. Biochem. Syst. Environ. Physiol.* 135(2):183–190.

McNulty, J. A. 1982. Effects of constant light and constant darkness on daily changes in the morphology of the pineal organ in the goldfish, *Carassius auratus. J. Neural. Transm.* 53(4):277–292.

Mezhnin, F. I. 1980. Evolution of the adrenal gland in the phylogeny of Cyclostomata and Pisces. *Vopr. Ikhtiol. (Journal of Ichthyology)* 20(4):679–693.

———. 1979. Stannius corpuscles of teleost fish. *Vopr. Ikhtiol. (Journal of Ichthyology)* 19(2):313–331.

———. 1979. Morphology and topography of interrenal and suprarenal glands of bony fishes. *Biol. Nauki (Moscow)* (3):33–38.

———. 1979. Interrenal and suprarenal glands of bony fishes. *Izv. Akad. Nauk SSSR ser. Biol.* (4):512–519.

Oguri, M. 1980. Presence of juxtaglomerular cells in trout kidneys. *Bull. Jpn. Soc. Sci. Fish.* 46(3):295–298.

———. 1980. A histological investigation of the juxtaglomerular cell granules in fish kidneys. *Bull. Jpn. Soc. Sci. Fish.* 46(7):797–800.

Pang, P. T. and A. Epple, eds. 1980. Evolution of vertebrate endocrine systems. *Graduate studies.* number 21, pp. 1–404. Texas Tech University Press, Lubbock, TX.

Tazawa, H., M. Mukai, and M. Ogawa. 1989. Immunohistochemical studies of juxtaglomerular cells and the corpuscles of Stannius in the eel, *Anguilla Japonica. Zoo. Sci.* 6(4):805–808.

Index

abdominal pore, 65
abdominis, 69, 73
abducens, 69, 73
abductors, 36, 37
acanthodians, 2
acanthopterygii, 3, 4
acousticofacialis, 73
acth, 83
actinopterygii, 1, 4, 26
adductor(s), 29, 31–34, 36, 39, 42
adductor mandibulae, 29, 31, 33, 34, 42
adenohypophysis, 81
afferent arteriole, 63
afferent branchial, 41, 52–55
afferent nerve, 74
agnatha, 1, 2
album centrale, 72
amia, 7, 42, 61, 79
amphibia, 2
amphibians, 83
amphicoelous, 21
ampulla(e), 76
anal, 1, 7, 9, 12, 23, 36, 43
anaspids, 11
anglerfishes, 3
anguilla, 7, 83
anguilliform, 7
anguilliformes, 3
angular, 15–17
anterior chamber, 78, 79
anus, 44, 46, 50, 65, 67
aorta, 52–55, 58–60, 64, 83
aortic radices, 55
aperture(s), 7–9, 79
appendage(s), 5, 8, 11, 39, 58
appendicular, 14, 23, 24, 36, 75
archinephric, 62, 64
artery
    afferent, 52, 53
    afferent arteriole, 63
    afferent branchial, 41, 54, 55
    aorta, 52–55, 58–60, 64, 83
    aortic radices, 55
    branchial, 41
    bulbus arteriosus, 44, 52–54
    carotid, 55, 61
    caudal mesenteric, 55, 56
    celiaco-mesenteric, 55, 56
    conus arteriosus, 52, 54
    duodenal, 50, 55, 56, 61
    efferent branchial, 41, 54, 55
    gastrohepatic, 43
    hepatic, 49
    intercostal, 55, 58
    internal carotid, 55
    pancreaticoduodenal, 61
    pneumatic, 55, 56
    posttrematic, 41
    pretrematic, 41
    renal, 65
    trematic, 54
    ventral aorta, 52–55, 83
articular, 15–17, 19, 22, 23, 31
atheriniformes, 3
atlas, 22, 61
atrioventricular, 54
atrium, 52–54, 60
aulopiformes, 3
auratus, 13, 65, 83
auricular, 70
aves, 2
axial, 14
axis, 5, 22, 36
axons, 71, 81

basibranchial, 20, 21, 83
basioccipital, 17, 18
basipterygia, 25
batracoidiformes, 3
beryciformes, 3
bichirs, 1
bile, 49, 50
bile duct, 49, 50
bladder, 11, 12, 43–45, 50, 57, 59, 61, 64–66, 76
blennies, 3
bone
    angular, 15–17
    articular, 15–17, 19, 22, 23, 31
    atlas, 22, 61
    axis, 5, 22, 36
    basioccipital, 17, 18
    basipterygia, 25
    centrale, 72
    ceratohyal, 19, 20, 35
    ceratohyals, 20, 34
    circumorbital, 14, 32
    cleithra, 24, 25, 54
    cleithrum, 15, 18, 24, 25, 35
    coracoid, 18, 24, 25, 36
    coracoids, 54
    dentary(ies), 8, 15–17, 31, 32, 34, 40
    dentigerous, 20, 41
    dermal, 14–17, 21, 23, 74, 76
    dermatocranium, 14, 24
    dermosphenotic, 32, 35
    ectopterygoid, 15, 18, 19, 32
    endochondral, 16, 17, 20, 21
    endocranium, 17, 26, 58
    entopterygoid, 19, 32
    epihyal, 19, 20

epiotic, 14, 15, 18, 19, 24, 35, 76
epurals, 23
ethmoid, 16, 17
exoccipital, 17, 32
frontal, 5, 14–17
hemal arch, 22, 23
hyoid, 18–21, 31, 34, 35, 41, 42, 73
hyomandibula, 31, 32
hypohyal, 20, 21, 35
hypurals, 23
interhyal, 19, 20
interopercular, 15, 17, 18, 31
interpterygiophore, 55, 58
lacrimal(s), 14–17
mandible, 8, 9, 31
mandibulae, 29, 31–34, 42
mandibular, 17, 29, 31, 34
maxillae, 8, 15–17
mesethmoid, 17
metapterygoid, 15, 18, 19, 31, 32
nasals, 17
neural arch, 21–23
neural spine, 21, 22, 24, 55
neurocranium, 14, 17, 31
occipital, 19, 73, 75
opercular, 15–19, 31–33, 35, 41, 42
operculi, 31, 32, 34, 42
operculum, 7, 9, 17, 19, 31, 32, 42
palatine, 16, 17, 19, 32, 41
parasphenoid, 17, 32, 35
parietal, 14–16, 18, 19, 43, 52, 63
parietals, 14
pectoral girdle, 15, 18, 24, 25, 31, 35
pelvic girdle, 25, 43
postcleithrum, 24, 25
postparietals, 14
posttemporal, 14, 15, 16, 18, 19, 24, 25, 32, 35
postzygapophyses, 22
premaxilla, 8, 15–18, 40
preoperculum, 19, 31
prezygapophysis, 17, 22
prootic, 19, 32, 76
pterotic, 15, 32, 35
pterygiophore, 21, 23
pterygiophores, 23, 36
pterygoid(s), 15, 17–19
quadrate, 15–17, 19, 31, 32
radial(s), 24–26, 36
scapula, 18, 24, 25, 36
sphenotic, 15, 16, 19
splanchnocranium, 14, 19
subopercular, 15, 17, 18, 35
supracleithrum, 15, 18, 24, 25
symplectic, 19, 31, 32
urohyal, 21, 35
urostyle, 21, 23, 24, 83
vertebrae, 1, 14, 17, 21–24, 26, 54, 73
vertebral, 21, 26, 27, 73
vomer, 16, 17, 19, 41
bonytongues, 3
bowfin, 3, 7, 42, 79
brachiopterygii, 1
brain
    aqueduct of sylvius, 70, 71
    bulbaris, 72
    facial lobe(s), 70, 71
    fibrosum et griseum, 72

forebrain, 68, 79
habenula(r), 70, 82, 83
hypothalamus, 68, 71, 80, 81
impar, 70, 71
inferiores, 71
medulla oblongata, 69, 70, 72, 73
mesencephalicus, 72
mesencephalon, 68–70, 72
metencephalon, 69
olfactory lobes, 68, 69
olfactory tract, 69
optic chiasma, 71
optic lobe, 69, 71, 72, 83
optic tectum, 70–73
rhombencephalon, 68, 69
subcommissural, 83
sylvius, aqueduct of, 70, 71
synapses, 71
tectum, 70–73, 83
tela choroidea, 70
telencephalon, 68–73
tori semicirculars, 70
tracts, 71–73
tuberculi impar, 70
valvula, 70, 71
valvular, 54
valvuli, 71
branchial (arteries), 18–20, 31, 41, 52, 54, 55, 83
branchiostegal, 20, 21, 31, 35, 41, 42
brasiliensis, 65
buccal, 31, 32, 34, 35, 42
bulbaris, 72
bulbus arteriosus, 44, 52–54
butterflyfish, 3

caeca, 50, 56, 57, 61, 82
calcitonin, 83
carangiform, 7
carassius, 13, 65, 83
cardiac, 54
cardinal, 52–54, 58–60
carotid, 55, 61
carpio, 65
cartilage
    branchiostegal, 20, 21, 31, 35, 41, 42
    ceratobranchial, 20, 21
    chondrocranium, 14
    epibranchial, 20, 21
    Meckel's, 16, 17, 31, 32
    pedicle, 78, 79
    pterygiophore(s), 21, 23, 36
caudal fin, 7–9, 12, 23, 36
caudal mesenteric artery, 55, 56
cavefish, 3
celiaco-mesenteric, 55, 56
cenozoic, 2
central canal, 71, 74, 82
centrale, 72
centrum, 21–23
cephalaspidomorphi, 1
ceratobranchial, 20, 21
ceratohyal(s), 19, 20, 34, 35
ceratotrichia, 23
cerebellaris, 72
cerebellum, 69–73, 79
characiformes, 3
chimaeras, 1

Chondrichthyes, 1, 2
chondrocranium, 14
chondrostei, 1
chordata, 1, 4
choroidea, 70
choroid layer, 79
ciclids, 3
ciliary, 79
circumorbital, 14, 32
cleithra, 24, 25, 54
cleithrum, 15, 18, 24, 25, 35
clingfishes, 3
cloaca, 50, 62, 64
clupeiformes, 3
clupeomorpha, 3
coelom, 43, 64
coelomic, 43, 64, 65
collecting tubule, 62, 63
commissures, 72
communicantes, 74
condyle, 15, 16
conjunctiva, 78
conus arteriosus, 52, 54
coracoid(s), 18, 24, 25, 36, 54
cornea, 78, 79
corpus, 46, 48–50, 70, 71
cortex, 71, 82
corticosterone, 82
corticotropin, 81
cranium, 17, 19, 73, 79, 83
cretaceous, 2
crossopterygian, 11
crossopterygii, 1, 26
cruciatus, 72
ctenii, 12, 13
ctenoid, 12, 13
cuboidal, 11, 45, 46, 48, 50
cupula, 76
cusps, 52, 54
cutaneous, 58, 83
cuvier, duct of, 52, 58, 60
cyclostomata, 84
cypriniformes, 3
cyprinodontiformes, 3
cyprinus, 65
cystic, 61
cytoplasm, 11, 50

dactylopteriformes, 3
dendrites, 71
dentary(ies), 8, 15–17, 31, 32, 34, 40
dentigerous, 20, 41
dermal, 14–17, 21, 23, 74, 76
dermatocranium, 14, 24
dermis, 11, 12, 14, 74
dermosphenotic, 32, 35
dermotrichia, 13
devonian, 2
diastole, 54
diencephalon, 68–71, 80, 83
dilator operculi, 32, 34, 42
diphycercal, 11, 12
dipneusti, 1
distalis, 81
diverticulum, 50
dorsal aorta, 52, 54, 55, 58–60, 64
dorsal fin, 8, 9, 12, 23, 27

dorsal root ganglion, 74
dragonfishes, 3
duct
    archinephric, 62, 64
    bile duct, 49, 50
    collecting tubule, 62, 63
    cystic, 61
    excretory, 62, 64
    oviduct, 65, 67
    pneumatic, 45
    sylvius, 70, 71
    vas deferens, 65, 66
ductules, 50
duodenum, 46, 48, 50, 55, 56, 61, 65

ectoderm, 11
ectopterygoid, 15, 18, 19, 32
eelpouts, 3
efferent (nerve), 74
efferent arteriole, 63
efferent branchial, 41, 54, 55
elasmobranch(s), 11, 58, 82
elasmobranchii, 1
elopiformes, 3
elopomorpha, 3
enamel, 13
endochondral, 16, 17, 20, 21
endocranium, 17, 26, 58
entopterygoid, 19, 32
epaxial, 23, 29, 31, 55, 74
epibranchial, 20, 21
epidermis, 11–13
epihyal, 19, 20
epiotic, 14, 15, 18, 19, 24, 35, 76
epiphysis, 83
epipleural, 21, 23
epithelium, 8, 45, 46, 48–50, 63, 73, 79
epurals, 23
esophagus, 40, 41, 46, 48, 51, 55, 57, 83
ethmoid, 16, 17
euphysoclists, 45
euteleostei, 3, 4
excretory, 62, 64
exoccipital, 17, 32
eye, 9, 19, 32, 35, 58, 68, 71, 73, 78, 79, 83

facial, 69–71, 73, 74
facial lobe, 70
falciform, 44
fasciculi, 72, 73
fibrosum et griseum, 72
filament(s), 13, 41, 42, 55
flavescens, 1, 3, 4, 65, 78
fluviatilis, 1, 4, 26, 39, 42, 51, 67, 76, 79, 80, 83
flyingfishes, 3
follicle, 67
forebrain, 68, 79
fossa, 83
frontal, 5, 14–17
fundic stomach, 43
fusiform, 7

gadiformes, 3
gall bladder, 50, 61
ganglia, 71, 74
gasterostiformes, 3
gastric, 48–51, 55–57, 61, 80

gastric artery, 55–57
gastric vein, 61
gastrohepatic ligament, 43
geniohyoideus, 32, 34, 35, 37, 42
geophagus, 65
gill raker(s), 40, 41
gill slit(s), 40–42
ginglymodi, 3
gland
    adenohypophysis, 81
    gonad, 55–57, 59
    gonads, 43, 55, 64, 82
    hypophysis, 71, 80
    langerhans, 50, 82
    leydig's, 83
    liver, 43, 44, 46, 49–53, 56–59
    ovary, 59, 64, 65, 67, 80
    pancreas, 43, 44, 46, 49–51, 56, 59, 61, 80, 82
    pars distalis, 81
    pineal, 68–70, 82, 83
    pituitary, 70, 71, 79–83
    Stannius corpuscle, 80, 82, 83
    testes, 59, 64–66, 80, 83
    thyroid, 81–83
    ultimobranchial, 83
glomeruli, 64, 65
glomerulus, 62–64, 67
glossopharyngeal, 69, 73–75
gnathostomata, 1, 4
gnathostomes, 6, 42
gobies, 3
gobiesociformes, 3
gonad, 55–57, 59
gonadal vein, 59
gonadotropin, 81
gonads, 43, 55, 64, 82
gonorynchiformes, 3
greeneyes, 3
griseum, 72
gurnards, 3
gymnarchus, 11
gymnotiformes, 3
gymnotus, 7

habenula, 70
habenular, 82, 83
hagfishes, 1
halecomorphi, 3
halecostomi, 3, 4
hatchetfishes, 3
hemal arch, 22, 23
hemibranchs, 41
hepatic, 49, 50, 52–55, 59–61
hepatic artery, 49
hepatic portal vein, 49–51, 59
hepatic sinus, 53, 59
hepatic vein, 54, 60
herbivorous, 46
heterocercal, 7, 11, 12
holobranch, 41
holocephalan, 67
holocephali, 1
holonephric, 64
holonephros, 64
homocercal, 11, 12

hormone
    acth, 83
    calcitonin, 83
    corticosterone, 82
    corticotropin, 81
    gonadotropin, 81
    hypocalcin, 82
    insulin, 50, 82
    melanotropin, 83
    oxytocin, 81
    prolactin, 81
    renin, 82
    secretin, 82
    thyrotropin, 81
    vasotocin, 81
hyohyoideus, 35, 42
hyoid, 18–21, 31, 34, 35, 41, 42, 73
hyoideus, 32, 34, 35, 42
hyomandibula(r), 15, 18–20, 26, 31, 32
hypaxial, 23, 29, 31, 35, 55, 74
hypobranchial, 20, 21
hypocalcin, 82
hypocercal, 11
hypohyal, 20, 21, 35
hypophysis, 71, 80
hypothalamus, 68, 71, 80, 81
hypurals, 23

impar, 70, 71
indostomiformes, 3
inferiores, 71
inferior oblique, 35, 38, 78
inferior rectus, 35, 38
infundibulum, 71, 80, 83
insulin, 50, 82
intercostal, 55, 58, 59
interhyal, 19, 20
intermandibularis, 31, 32, 34, 37
intermedia, 81, 83
internal carotid, 55
interopercular, 15, 17, 18, 31
interpterygiophore, 55, 58
interrenal, 81–83
interspinal, 55, 58
intestine(s), 43, 44, 46, 48–51, 55, 59, 61, 65, 82
invagination, 64
irideus, 13
iris, 78, 79

joint
    symphysis, 15, 17, 31, 32
jugular, 59

kidney, 44, 54, 56, 57, 59–65, 67, 82, 83
kidneys, 43, 45, 58, 59, 62, 64, 67, 83
kinocilium, 74, 76
knifefishes, 3

lacrimal(s), 14–17
lagena, 77–79
lamellae, 42
lampriformes, 3
langerhans, islets of, 50, 82
lanternfishes, 3
lateral line, 9, 43, 58, 68, 70, 73, 74, 76

lateral rectus, 35, 38
lens, 78, 79
lepidotrichia, 21, 23, 36
levator arcus palatini, 29, 32, 42
Leydig cells, 83
ligament
    falciform, 44
    gastrohepatic, 43
    maxillo-mandibular, 29, 31, 34
    suspensory, 78, 79
lightfishes, 3
livebearers, 3
liver, 43, 44, 46, 49–53, 56–59
lizardfishes, 3
lobe, 69, 71, 72, 83
lophiiformes, 3
lumen, 46, 48, 50, 67, 83
lungfishes, 1

mammalia, 2
mandible, 8, 9, 31
mandibulae, 29, 31–34, 42
mandibular, 17, 29, 31, 34
maxillae, 8, 15–17
maxillo-mandibular ligament, 29, 31, 34
Meckel's cartilage, 16, 17, 31, 32
medulla oblongata, 69, 70, 72, 73
melanin, 83
melanotropin, 83
membrane
    choroidea, 70
    choroid layer, 79
    falciform, 44
    mesenteries, 43, 52
    mesentery, 43, 48, 50
    mesovarium, 67
    parietal peritoneum, 43, 52
    pericardial, 43, 52–54, 83
    pericardium, 52, 54
    peritoneal, 43, 52
    pleural, 21, 23
    pleuroperitoneal, 23
    serosa, 48
    serous, 43
    tela choroidea, 70
    transverse septum, 31, 52
    tympanum, 19
    ventral mesentery, 43
    visceral peritoneum, 43
mesencephalicus, 72
mesencephalon, 68–70, 72
mesenchymal, 64
mesenteries, 43, 52
mesentery, 43, 48, 50
mesethmoid, 17
mesoderm, 11
mesonephric, 64
mesonephros, 64
mesovarium, 67
metapterygoid, 15, 18, 19, 31, 32
metencephalon, 69
milkfish, 3
molas, 3
mooneyes, 3
mouth, 7–9, 19, 31, 32, 40–42, 45, 48, 71

mucosal epithelium, 46
mucus, 11, 12, 48, 50
muscle
    abductor(s), 36, 37
    adductor(s), 29, 31–34, 36, 39, 42
    adductor mandibulae, 29, 31, 33, 34, 42
    appendicular, 14, 23, 24, 36, 75
    dilator operculi, 32, 34, 42
    epaxial, 23, 29, 31, 55, 74
    geniohyoideus, 32, 34, 35, 37, 42
    hyohyoideus, 35, 42
    hyoideus, 32, 34, 35, 42
    hypaxial, 23, 29, 31, 35, 55, 74
    hypobranchial, 20, 21
    inferiores, 71
    inferior oblique, 35, 38, 78
    inferior rectus, 35, 38
    intermandibularis, 31, 32, 34, 37
    lateral rectus, 35, 38
    levator arcus palatini, 29, 32, 42
    muscularis externa, 50
    myofibrils, 30
    myomere, 30, 38, 39
    myomeres, 36, 58
    myotome, 27, 28, 31
    myotomes, 28, 31, 39, 79
    pharyngobranchial, 21
    rectus abdominis, 36
    sphincter, 48
    sternohyoideus, 35–37, 42, 43
    suborbital, 14, 15, 74
    subvertebral, 45
    superficialis, 72
    superior oblique, 35, 38, 73, 77
    superior rectus, 35, 38, 77
muscularis externa, 50
myctophiformes, 3
myelencephalon, 69, 71, 72
myelin, 71
myofibrils, 30
myomere(s), 30, 36, 38, 39, 58
myosept(a, um), 12, 23, 27, 28, 30, 34, 35, 39
myotome(s), 27, 28, 31, 39, 79
myxini, 1

nasal(s), 7–9, 14–17, 73, 79
neopterygii, 1, 4
nephrostomes, 64
nerve, 28, 35, 38, 48, 68, 70–79, 81
    abducens, 69, 73
    acousticofacialis, 73
    afferent, 41, 52–55, 63, 74
    axons, 71, 81
    dendrites, 71
    dorsal root, 74
    facial, 69, 70, 71, 73, 74
    ganglia, 71, 74
    glossopharyngeal, 69, 73–75
    myelin, 71
    nervus terminalis, 73
    neurons, 54, 80
    oculomotor, 38, 69, 73
    olfactory, 8, 68, 69, 71, 73, 79
    optic, 35, 38, 68–73, 77–79, 83
    otic, 19, 24

spinal, 28, 68, 73, 74, 75
sympathetic, 54, 68, 73, 74, 82
synapses, 71
tracts, 71–73
trigeminal, 69, 73
trochlear, 38, 69, 72, 73
vagal, 71
vagus, 69, 73–75
ventral root, 74
nervus terminalis, 73
neural, 21–24, 55, 73, 74, 76, 79–81, 83
neural arch, 21–23
neural spine, 21, 22, 24, 55
neurocranium, 14, 17, 31
neuromasts, 74, 76, 77
neurons, 54, 80
notacanthiformes, 3
notochord, 21
notochordal canal, 21, 22
nuclei, 71, 72, 80, 81

oblique, 35, 38, 73, 77, 78
occipital, 19, 73, 75
oculomotor, 38, 69, 73
olfactory, 8, 68, 69, 71, 73, 79
olfactory epithelium, 8, 73, 79
olfactory lobes, 68, 69
olfactory tract, 69
opercular, 15–19, 31–33, 35, 41, 42
operculi, 31, 32, 34, 42
operculum, 7, 9, 17, 19, 31, 32, 42
ophidiiformes, 3
opisthonephric, 64, 82
opisthonephroi, 64
opisthonephros, 64
opisthotic, 18, 19, 76
optic, 35, 38, 68–73, 77–79, 83
optic chiasma, 71
optic lobe, 69, 71, 72, 83
optic nerve, 35, 38, 70, 71, 77–79
optic tectum, 70–73
oral, 40–42, 80
ordovician, 2
ostariophysi, 3
osteichthyan, 13
osteichthyes, 1, 2, 4, 26
osteoglossiformes, 3
osteoglossomorpha, 3
ostracoderms, 2, 11, 14
otic, 19, 24
ovary, 59, 64, 65, 67, 80
oviduct, 65, 67
oxytocin, 81

palatine, 16, 17, 19, 32, 41
pancreas, 43, 44, 46, 49–51, 56, 59, 61, 80, 82
pancreaticoduodenal, 61
papilla(e), 64–66
paracanthopterygii, 3
paradoxus, 3
paraphysis, 70
parapineal, 70
parasphenoid, 17, 32, 35
parietal, 14–16, 18, 19, 43, 52, 63
parietals, 14
pars distalis, 81
pectoral fin, 9, 12, 21, 25, 36, 37, 43

pectoral girdle, 15, 18, 24, 25, 31, 35
pedicle, 78, 79
peduncle, 7, 9
pegasiformes, 3
pelvic, 1, 4, 9, 11, 12, 25, 26, 36, 37, 43, 58, 73, 75
pelvic fin, 4, 9, 26, 36, 37
pelvic girdle, 25, 43
Perca, 1, 3, 4, 26, 39, 42, 51, 65, 67, 76, 78–80, 83
percidae, 1, 4, 26
percids, 1
perciforme(s), 1, 3, 4, 39
percinae, 4
percini, 4
percoidea, 4
percopsiformes, 3
pericardi(al, um), 43, 52–54, 83
peritone(al, um), 43, 52, 65
permian, 2
pharyngeal, 39, 41
pharyngobranchial, 21
pharynx, 40–42, 45, 71
phylogenetic, 1, 2
physoclists, 45
physostomes, 45
pineal, 68–70, 82, 83
pipefishes, 3
pituitary, 70, 71, 79–83
placoderma, 2
placoid, 13
plasma, 82
pleural, 21, 23
pleuronectiformes, 3
pleuroperitoneal, 23
plexus, 73, 75
plica, 72
pneumatic, 45, 55, 56
poikilotherms, 31
pore, 64–67
portal, 49–52, 58–61, 64, 81, 83
postcleithrum, 24, 25
posterior chamber, 78, 79
postparietals, 14
posttemporal, 14–16, 18, 19, 24, 25, 32, 35
posttrematic, 41
postzygapophyses, 22
premaxilla, 8, 15–18, 40
preoperculum, 19, 31
pretetrapod, 73
pretrematic, 41
prezygapophysis, 17, 22
process, 15–17, 19, 22, 32, 35, 64
    condyle, 15, 16
    epiphysis, 83
    epipleural, 21, 23
    postzygapophyses, 22
    prezygapophysis, 17, 22
    spine(s), 1, 12–14, 17, 21–25, 55
    supraoccipital, 14–19
    supraorbital, 14
    tuberculum, 71
    urohyal, 21, 35
    urostyle, 21, 23, 24, 83
    zygapophyses, 22
prolactin, 81
pronephr(ic, os), 64
prootic, 19, 32, 76
propria, 46, 48

proprius, 35
protacanthopterygii, 3
pseudobranchs, 41
pterotic, 15, 32, 35
pterygiophore(s), 21, 23, 36
pterygoid(s), 15, 17–19
pubic, 26
pupil, 78, 79
pylor(ic, us), 46, 48–50, 56, 57, 61, 82

quadrate, 15–17, 19, 31, 32
quadrigeminum, 71

radial(s), 24–26, 36
raker(s), 41
rami communicantes, 74
rayfish, 11
rectilinear, 50
rectum, 44, 46, 50
rectus, 35, 36, 38, 72, 73, 77
rectus abdominis, 36
renal, 52, 59, 60, 63–65, 67, 83
renal artery, 65
renal portal, 52, 59, 60, 64, 83
renal portal vein, 60
renal tubule, 67
renal vein, 65
renin, 82
reptilia, 2
rete (mirabile), 55, 64
retina, 73, 78, 79
retroperitoneal, 43
rhombencephalon, 68, 69
ribbonfishes, 3
rostrum, 58

saccul(e, us), 76–79, 82, 83
saccus dorsalis, 70
saccus vasculosus, 69–72
sagitta, 77
sagittal, 5, 53, 78
salmoniformes, 3
scapula, 18, 24, 25, 36
sclera, 78, 79
scopelomorpha, 3
scorpaeniformes, 3
sculpins, 3
seahorses, 3
seamoths, 3
secretin, 82
semicircular canal(s), 76–78
semicirculari(s), 69–72
sense organ
    abdominal pore, 65
    lateral line, 9, 43, 58, 68, 70, 73, 74, 76
    lens, 78, 79
    melanin, 83
    neuromasts, 74, 76, 77
    olfactory, 8, 68, 69, 71, 73, 79
    olfactory epithelium, 8, 73, 79
    sclera, 78, 79
    semicircular canal(s), 76–78
    semicirculari, 71, 72
    sinoatrial, 54
    vitreous body, 78, 79
sensory, 71–74, 79
septum, 23, 31, 52, 55

serosa, 48
serous, 43
sertoli cells, 83
silurian, 2
siluriformes, 3
sinoatrial, 54
sinus, 44, 52–54, 56–60, 64, 65, 83
sinus venosus, 44, 52–54, 56–60, 83
siphonal, 50
skeleton
    amphicoelous, 21
    anaspids, 11
    angular, 15–17
    appendage, 5
    appendages, 5, 8, 11, 39, 58
    articular, 15–17, 19, 22, 23, 31
    atlas, 22, 61
    axis, 5, 22, 36
    basibranchial, 20, 21, 83
    basipterygia, 25
    centrale, 72
    centrum, 21–23
    ceratobranchial, 20, 21
    ceratohyal, 19, 20, 35
    ceratohyals, 20, 34
    ceratotrichia, 23
    chondrocranium, 14
    circumorbital, 14, 32
    cleithr(a, um), 15, 18, 24, 25, 35, 54
    condyle, 15, 16
    coracoid(s), 18, 24, 25, 36, 54
    cranium, 17, 19, 73, 79, 83
    dentary, 8, 15–17, 31, 32, 34, 40
    dentigerous, 20, 41
    dermal, 14–17, 21, 23, 74, 76
    dermatocranium, 14, 24
    diphycercal, 11, 12
    ectopterygoid, 15, 18, 19, 32
    endochondral, 16, 17, 20, 21
    endocranium, 17, 26, 58
    entopterygoid, 19, 32
    epibranchial, 20, 21
    epihyal, 19, 20
    epiotic, 14, 15, 18, 19, 24, 35, 76
    epiphysis, 83
    epipleural, 21, 23
    epurals, 23
    ethmoid, 16, 17
    exoccipital, 17, 32
    fossa, 83
    frontal, 5, 14–17
    hemal arch, 22, 23
    homocercal, 11, 12
    hyoid, 18–21, 31, 34, 35, 41, 42, 73
    hyomandibula(r), 15, 18–20, 26, 31, 32
    hypohyal, 20, 21, 35
    hypurals, 23
    interhyal, 19, 20
    interopercular, 15, 17, 18, 31
    interpterygiophore, 55, 58
    lacrimal(s), 14–17
    lepidotrichia, 21, 23, 36
    mandible, 8, 9, 31
    mandibulae, 29, 31–34, 42
    mandibular, 17, 29, 31, 34
    maxillae, 8, 15–17
    Meckel's cartilage, 16, 17, 31, 32

mesethmoid, 17
metapterygoid, 15, 18, 19, 31, 32
nasal capsule, 14
nasals, 17
neural arch, 21–23
neural spine, 21, 22, 24, 55
neurocranium, 14, 17, 31
notochord, 21
notochordal canal, 21, 22
occipital, 19, 73, 75
opercular, 15–19, 31–33, 35, 41, 42
opercul(i, um), 7, 9, 17, 19, 31, 32, 34, 42
palatine, 16, 17, 19, 32, 41
parasphenoid, 17, 32, 35
parietal(s), 14–16, 18, 19, 43, 52, 63
pectoral girdle, 15, 18, 24, 25, 31, 35
pedicle, 78, 79
pelvic girdle, 25, 43
pneumatic, 45, 55, 56
postcleithrum, 24, 25
postparietals, 14
posttemporal, 14–16, 18, 19, 24, 25, 32, 35
postzygapophyses, 22
premaxilla, 8, 15–18, 40
preoperculum, 19, 31
prezygapophysis, 17, 22
prootic, 19, 32, 76
pterotic, 15, 32, 35
pterygiophore, 21, 23
pterygiophores, 23, 36
pterygoid(s), 15, 17, 18, 19
pubic, 26
quadrate, 15–17, 19, 31, 32
radial(s), 24–26, 36
rostrum, 58
scapula, 18, 24, 25, 36
sphenotic, 15, 16, 19
spine(s), 1, 12–14, 17, 21–25, 55
splanchnocranium, 14, 19
subopercular, 15, 17, 18, 35
supracleithrum, 15, 18, 24, 25
supraoccipital, 14–19
supraorbital, 14
symphysis, 15, 17, 31, 32
symplectic, 19, 31, 32
tuberculum, 71
urohyal, 21, 35
urostyle, 21, 23, 24, 83
vertebra(e, l), 1, 14, 17, 21–24, 26, 27, 54, 73
vertebrata, 4, 6, 42
vomer, 16, 17, 19, 41
zygapophyses, 22
somatic, 55, 71, 73, 80
spermatic, 65
sphenotic, 15, 16, 19
sphincter, 48
spinal nerve(s), 28, 68, 73, 74, 75
spine(s), 1, 12–14, 17, 21–25, 55
spiracle, 41
splanchnocranium, 14, 19
spleen, 43, 46, 55, 56, 59, 61, 64, 83
splenic, 55, 61
squamous, 11

squirrelfishes, 3
Stannius corpuscle, 80, 82, 83
stenopterygii, 3
stereocilia, 74, 76
sternohyoideus, 35–37, 42, 43
sticklebacks, 3
stizostedion, 1
stomach, 43, 46, 48–51, 55–57, 59, 61, 65
stomiiformes, 3
stratum germinativum, 11
subcarangiform, 7
subcommissural, 83
subcutis, 11, 12
submucosa, 46, 48, 49
subopercular, 15, 17, 18, 35
suborbital, 14, 15, 74
subvertebral, 45
superficialis, 72
superior oblique, 35, 38, 73, 77
superior rectus, 35, 38, 77
supracleithrum, 15, 18, 24, 25
supraoccipital, 14–19
supraorbital, 14
suspensory ligament, 78, 79
sylvius, aquaduct of, 70, 71
sympathetic nerve, 54, 68, 73, 74, 82
symphysis, 15, 17, 31, 32
symplectic, 19, 31, 32
synapses, 71
synbranchiformes, 3
systole, 54

tectum, 70–73, 83
tela choroidea, 70
telencephalon, 68–73
teleost(s), 1, 3, 11, 13, 23, 26, 39, 42, 45, 51, 54, 59, 61, 64, 65, 73, 79, 82, 83
teleostei, 1, 3, 4
testes, 59, 64–66, 80, 83
tetraodontiformes, 3
tetrapods, 11, 19, 22, 24, 26, 27, 39, 70
thoracic, 1, 12
thyroid, 81–83
thyrotropin, 81
toadfishes, 3
tori semicircularis, 70
tracts, 71–73
transverse septum, 31, 52
trematic, 54
triassic, 2
trigeminal, 69, 73
trochlear, 38, 69, 72, 73
tuberculi impar, 70
tuberculum, 71
tympanum, 19

ultimobranchial, 83
urinary bladder, 64, 65
urinary papilla, 64–66
urogenital, 7, 12, 23, 62, 64, 65
urohyal, 21, 35
urophysial, 83
urophysis, 83
urostyle, 21, 23, 24, 83
utricle, 76, 77

vagal, 71
vagus, 69, 73–75
valvula, 70, 71
valvular, 54
valvuli, 71
vasculosus, 68–72
vas deferens, 65, 66
vasotocin, 81
vein, 38, 48–51, 54, 57–62, 65
    cardinal, 52–54, 58–60
    cuvier, 52, 58, 60
    gonadal, 59
    hepatic, 54, 60
    hepatic portal, 49–51, 59
    hepatic sinus, 53, 59
    intercostal, 59
    jugular, 59
    portal, 49–52, 58–61, 64, 81, 83
    renal, 65
    renal portal, 52, 59, 60, 64, 83
    sinus venosus, 44, 52–54, 56–60, 83
ventral aorta, 52–55, 83
ventral mesentery, 43
ventral root, 74
ventricle, 44, 52–54, 56, 57, 60, 70–72, 83
vertebra(e,l), 1, 14, 17, 21–24, 26, 27, 54, 73
vertebrata, 4, 6, 42
vesicle, 64, 83
vestibule, 40
villi, 50
visceral, 43, 44, 52, 55–57, 63, 82
visceral peritoneum, 43
vitreous body, 78, 79
vomer, 16, 17, 19, 41

zeiformes, 3
zygapophyses, 22